I0034993

CRYSTAL CHEMISTRY

From Basics to Tools for Materials Creation

CRYSTAL CHEMISTRY

From Basics to Tools for Materials Creation

Gérard Férey

University of Versailles, France

World Scientific

NEW JERSEY · LONDON · SINGAPORE · BEIJING · SHANGHAI · HONG KONG · TAIPEI · CHENNAI · TOKYO

Published by

World Scientific Publishing Co. Pte. Ltd.

5 Toh Tuck Link, Singapore 596224

USA office: 27 Warren Street, Suite 401-402, Hackensack, NJ 07601

UK office: 57 Shelton Street, Covent Garden, London WC2H 9HE

Library of Congress Cataloging-in-Publication Data
Names: Ferey, Gerard, 1941–
Title: Crystal chemistry : from basics to tools for materials creation /
    Gerard Ferey, University of Versailles, France.
Description: New Jersey : World Scientific, 2016. | Includes bibliographical references.
Identifiers: LCCN 2016039522| ISBN 9789813144187 (hardcover) | ISBN 9789813144194 (pbk.)
Subjects: LCSH: Crystallography. | Solid state chemistry.
Classification: LCC QD905.2 .F47 2016 | DDC 548/.3--dc23
LC record available at https://lccn.loc.gov/2016039522

British Library Cataloguing-in-Publication Data
A catalogue record for this book is available from the British Library.

Copyright © 2017 by World Scientific Publishing Co. Pte. Ltd.

All rights reserved. This book, or parts thereof, may not be reproduced in any form or by any means, electronic or mechanical, including photocopying, recording or any information storage and retrieval system now known or to be invented, without written permission from the publisher.

For photocopying of material in this volume, please pay a copying fee through the Copyright Clearance Center, Inc., 222 Rosewood Drive, Danvers, MA 01923, USA. In this case permission to photocopy is not required from the publisher.

Typeset by Stallion Press
Email: enquiries@stallionpress.com

To Jean Pannetier

Preface

In the 21st century, a chemist must know how to exploit the structures of the solids he discovers. He must extract the essential information which will allow him to not only see how the atoms are arranged in these solids but, far beyond, to anticipate, at least qualitatively, their expected physical properties.

"Reading" a structure is not as easy as it seems. It is a long frequentation of the nano-world, the sum of diverse observations leading to understand, at the end, the organization of matter. It is the apprenticeship of an extreme simplification which, by the way, facilitates the memorization of structures. This being done, new ideas of synthesis, and sometimes predictions of the structure of new solids emerge from this careful examination which becomes also a source of inspiration. Some examples will be given as we proceed along this book.

In it, two words will appear frequently: crystallography and crystal chemistry. They are close but consecutive. Many people think that crystallography is relevant of physics and crystal chemistry, of course, of chemistry. Historically, it is true, but things are changing. The crystallographer was primitively the physicist who, starting from a crystal or a powder, was able — among other tasks — to determine from X-ray, neutron or electron diffraction the repartition of atoms in the three-dimensional space. This often led, besides a drawing, to the publication of a table of reduced atomic coordinates within a cell, the symmetry and dimensions of which were being determined in the first steps of the study. The work of the crystal chemist began after, but time has changed such habits. Now, due to both important technological developments and the onset of interdisciplinarity with time, chemists have become crystallography users and are now able to determine themselves the structures, whereas physicists go further in their physical knowledge of matter.

What is crystal chemistry? It is first the way to describe in the easiest manner the structural arrangements deduced from the crystallographic study. Further, these simple descriptions allow to establish useful comparisons with other structural arrangements, to deduce some relations, sometimes unexpected between them and, consequently, provide classifications within the structures of the solid matter and rules. Once understood the rules which govern the atomic arrangements of solids, this can serve to imagine new organizations of the matter, hitherto unknown, could be virtual at the very beginning, but an intelligent chemistry can then render them real. Finally, due to the progresses of the relations between solid state physics and chemistry, a good crystal chemist of the 21st century, looking at a given structure, is able to qualitatively predict the physical properties of the corresponding solid and undertake quantitative measurements of these properties.

The largest part of my scientific career has been devoted to this fascinating approach of the solid which, well understood, allows, for the chemist that I am, the creation of numerous new solids with, sometimes, unprecedented properties. I am also a teacher, interested to share my passion with new generations. I did that all my life. Unfortunately, at variance to crystallography for which numerous excellent books are available for helping students, a textbook accessible to the largest audience does not currently exist. Those which exist start at a very high level and are reserved for educated researchers. For a colleague in charge of teaching crystal chemistry, like me, the current evolution of the scientific knowledge of undergraduates, particularly in mathematics, leads to the initial reaction of a majority of students whose comments are: "more maths!" or "I am not good in geometry! Mainly where 3D space is concerned!", or even "I do not know how to draw"… All sentences reflecting the rejection or apprehension of the students in chemistry at the beginning of the course…

To convince them that crystal chemistry is not as irksome as they think, it is first necessary to reassure them that a high school level mathematics is sufficient, mainly avoid the strict academic approach that is usually associated with accumulation of knowledge. This book chooses a more light approach, progressively transiting from qualitative to quantitative, taking also into account History and Arts related to Science. This book is the result of this approach. It must be better seen as a free trip amongst shapes rather than a textbook teaching rapidly what is essential. Therefore, particular attention was paid to quality of the drawings. As a result, the presentation of this book is rather unconventional, insisting more on the beauty of the forms than on mathematical developments. However, my long experience permits me to state that this approach, an alternative one on solid matter, was fruitful for a large number of my students, and even generated numerous rational tropisms toward the study of solids and materials.

Another eye? Indeed, the apprenticeship of a crystal chemist is first a training of the eyes which, by curiosity, tries to scrutinize in different ways the atomic arrangements, searching among the different ways of description of those (the most simple!) which highlight unexpected relations between apparently different structures, and from which classifications can emerge. In this sense, crystal chemistry is a description, and also — and mainly! — for the chemists who were able to decipher the codes, an outstanding source of creativity leading to, thermodynamics permitting, a great number of absolutely new and predicted architectures, bearing interesting properties for applications that can satisfy the concerns of society in energy, environment and health.

This book is not a high-level book dedicated to experts. Others have done so, brilliantly, before me, and their names will appear within the text. The book aims to be useful for undergraduates, to high schools teachers, and to those of my colleagues who complain about the absence of such textbooks, serving both as references for ensuring the basic knowledge and also for providing new approaches of the different facets of the solid state. I hope that this book will help all of them.

Intending to help, and owing to the particular significance of figures, I considered those colleagues who teach crystal chemistry. As their capacities as illustrators are eminently variable, I chose to add to this book a supplementary material, accessible at the following address: http://www.worldscientific.com/worldsci-books/10.1142/10144. It gathers in a PowerPoint format all the figures which are in the book. The users can easily extract them and project them during their lectures.

I wrote this book to celebrate the memory of my close friend Jean Pannetier. He passed away too early, some years ago. We had the same passion for crystal chemistry. Without a doubt, if he was still alive, we would have produced this book together with four hands.

This book is also a tribute to the pioneers who paved the way: Linus Pauling, the double Nobel Prize winner, who was the first to represent the atomic arrangements by assemblies of polyhedra instead of linked spheres; David Wadsley, from Melbourne (Australia), who developed electron microscopy for the visualization and elucidation of some mysteries of the arrangements in complex solids; Alexander Wells, and his famous books describing the chemical structures... A special mention to the quatuor composed of the Australian Bruce H. Hyde (†), the Swedish Sten Andersson, the Anglo-American Michael O'Keeffe and the French Jean Galy who, during the seventies, had already understood the organization of matter. They definitively inspired the young researcher that I was. Most of them also honored me with their friendship. They were too advanced in time and some of them did not receive the recognition they deserved. This book is a tribute to their lessons.

Finally, my friend and great crystal chemist Prof. Maryvonne Hervieu, also an international expert in the transmission electron microscopy of oxides, spent a lot of time to track down the numerous errors contained in the original manuscript and to suggest pertinent modifications. Without her, the quality of this book would not have been the same. She will find here the expression of my deep and friendly gratitude.

Gérard Férey
Paris, March 30, 2016.

Contents

Historical Introduction

Even if traces of archeo-geometry were found in the Sumerian and Babylonian civilizations before 3,000 years B.C., it is in Egypt that geometry and its applications really began with Imothep, the first known architect (2,700 B.C.) who built the Djoser pyramid in Saqqarah. It is only 600 years later that the Gizeh pyramids were edificated.

However, it is in Greece that geometry obtained with time its most significant developments. The names of their authors remain famous even now, but one often forgets both the chronology and the duration of the evolution of this knowledge.

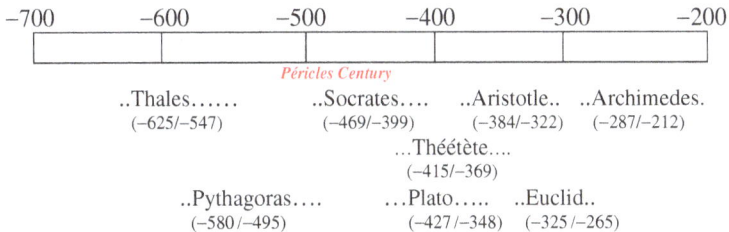

−700	−600	−500	−400	−300	−200

Péricles Century

..Thales......
(−625/−547)

..Socrates....
(−469/−399)

..Aristotle..
(−384/−322)

..Archimedes.
(−287/−212)

...Théétète....
(−415/−369)

..Pythagoras....
(−580/−495)

...Plato.....
(−427/−348)

..Euclid..
(−325/−265)

By the way, it is worthy to note that these famous names were, for most of them, philosophers seeking for **Harmony, Beauty and Purity**. The shapes were one of their manifestations. Thales, one of the seven Wise Men of Ancient Greece, also a merchant, a philosopher and an universal scholar at that time, was probably the first Greek to be inspired by geometry, after his long trips to Egypt and in Minor Asia. As a scholar, his audience was immense and caught the attention of the young Pythagoras. Even if there is not any written trace of their exchanges, the continuous oral transmission by their students and biographies written centuries after, assorted with charming legends, allowed their reputation to spread to us. Anyhow, geometry was born, primitively as an expression of the universal harmony.

The word *geometry* has of course a Greek origin: γεωμετρία (*geo*: earth; *metria*: measurement). The word *geo* also implies the notion of shapes, those which exist on earth. Intuitively, through the Sun and Moon, the first shape apprehended by humans was of course the sphere, considered as the purest shape, related to God, and therefore to Harmony. This was also developed by the post-Pythagorean school of Philolaos of Crotonia (–485/–390), a former student of Pythagoras and his transcriptor after his death. He developed the theory of the Harmony of Spheres in his book *About Nature*, in which it was postulated that the Universe was governed by harmonious numerical ratios, as in the case, for instance, of the distances between the planets. The legend says that Plato himself bought this book which had a strong impact over the centuries. It served later as a reference that was cited in the famous books of Plato *The Republic*, of Aristotle *About sky* and even by Copernic in 1543, in a letter to the Pope Paul III, who recognized his tribute to Philolaos: "Starting from there, I began, me also, to think about the mobility of the Earth"…

The future showed that the sphere was not the only shape interesting to philosophers. If the sphere represented the plenitude, one of the induced aspects of Harmony was also the regularity, the order. Nobody can tell who was the originator of this idea, and all sorts of hypotheses can be imagined … I have one, perfectly iconoclast, without any historical basis. It came to me during one of my visits to the Vatican Museum, in front of the famous painting of Raphael: *The School of Athens* (Fig. 1).

Fig. 1.   The painting *The School of Athens* by Raphael Sanzio (Vatican Museum).

In this superb tribute to both Greek philosophy and the most famous artists of the *Renaissance* period, two men are discussing in the center of the painting: Plato (represented with the face of Leonardo da Vinci) shows the sky to his disciple Aristotle. What did they say? Of course, nobody knows but one is free to imagine that, for instance, Plato informs Aristotle that, during his walks around Athens, he discovered perfectly cubic crystals of pyrite, these crystals, the faces of which being identical, and their edge lengths perfectly equal … Is it not a supplementary symbol of Harmony? Is it rare or frequent? One can also imagine that, after this discussion, Plato submitted the question to his disciples at his Academia. Whatever the legends, one thing is for sure: they showed that five polyhedra, and only five (the Platonic polyhedra), exhibit the double characteristics of having all their faces identical and their edges equal in length: the tetrahedron, the cube, the octahedron, the pentagonal dodecahedron and the icosahedron (Fig. 2).

Moreover, in the philosophical vision of the Universe, this scarcity ought to have a cosmological origin. For Plato (according to some, but from Timaeus de Locri, for others), these polyhedra were the symbols of what was considered in this period as the four elements: the tetrahedron representing the Fire, the cube the Earth, the octahedron the Air, the icosahedron the Water, the dodecahedron being the symbol of the Universe itself. What Plato[1] did not suspect was that he had inadvertently become with millenaries in advance, the first crystal chemist!…

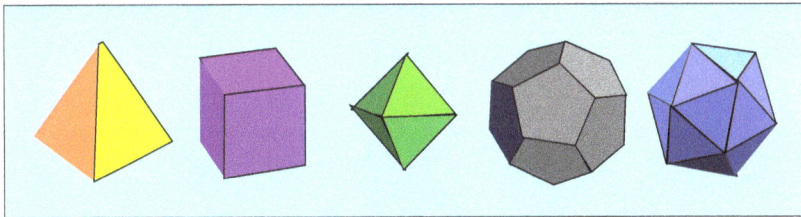

Fig. 2. The five Platonic polyhedra. (Left to right) The tetrahedron (Fire), the cube (Earth), the octahedron (Air), the pentagonal dodecahedron (Universe) and the icosahedron (Water).

---

[1]Here also, the legend is in favor of rich people… It prefered Plato. However, even if the name of Théétète is less known than Plato (but Plato cited him!), it must not be forgotten that, during the same period, this young mathematician (−415/−369) had already produced remarkable contributions on octahedra and icosahedra. Euclide, first, and centuries later, Kepler, Descartes and Euler were inspired by his achievements.

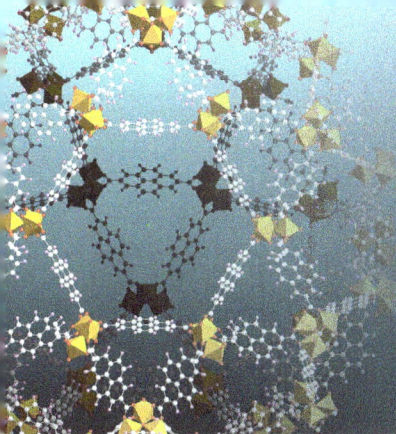

Chapter 1

A First Familiarization with Geometric Shapes: The Dense Packings

They have eyes, but don't see...
The Koran *(Surah 7, Verse 179)*

Why such a chapter in which neither crystallography nor chemistry appear? The notions presented here, which have the origins in both philosophy and mathematics, first concern the macroscopic Universe, but have also found a marvellous resonance in the nano-world of atomic arrangements as soon as the developments in physics at the beginning of 20th century paved way for this relation. It is therefore through these strange resonances that we shall progressively tackle crystal chemistry.

1.1 The Archimedian Polyhedra

The five Platonic polyhedra reflected the regularity and the identity of their faces. A question then arose: can other polyhedra present different kinds of faces, all of them remaining regular? More than a century after the death of Plato, Archimedes of Syracuse (–287/–212), a disciple of Euclid (–330?/–260?), provided the solution by introducing the concept of truncation, first applied to the Platonic polyhedra, and leading to the so-called Archimedian polyhedra, with different regular faces having edges of same length.

Figure 1.1 shows the progressive truncations of the vertices of the cube. They generate successively three Archimedian polyhedra: (i) the truncated cube with six regular octogonal faces and eight equilateral triangles, obtained when the truncation arises at one-third of the length of the edge; (ii) when done at the middle of each edge, it leads to the cuboctahedron, with six square faces and eight equilateral triangles; when it reaches two-thirds of the edge, the truncated octahedon appears (six square and eight regular hexagonal faces). Finally, a truncation corresponding to a full edge gives the octahedron, another Platonic polyhedron with its eight triangular faces.

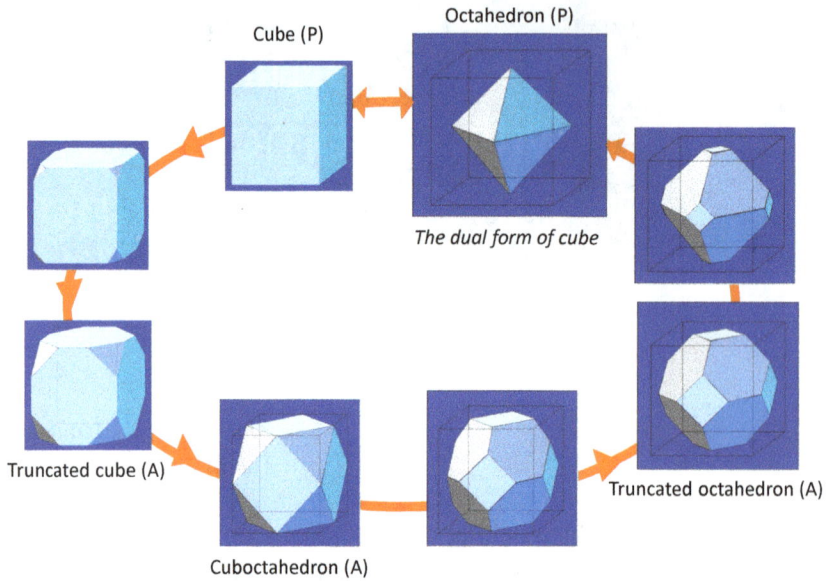

Fig. 1.1. Construction of three Archimedian polyhedra (noted A) [truncated cube, cuboctahedron and truncated octahedron], and of the Platonic octahedron (P), by progressive truncations of the vertices of a cube.

The octahedron is the *dual form* of the cube. Duality represents the reciprocity between two polyhedra. It can be illustrated when considering the center of a polyhedron and the lines perpendicular to each face containing this center. Joining together the intersections of these lines with the faces of the original polyhedron provides the shape of the dual polyhedron (Fig. 1.2 for the duality cube-polyhedron).

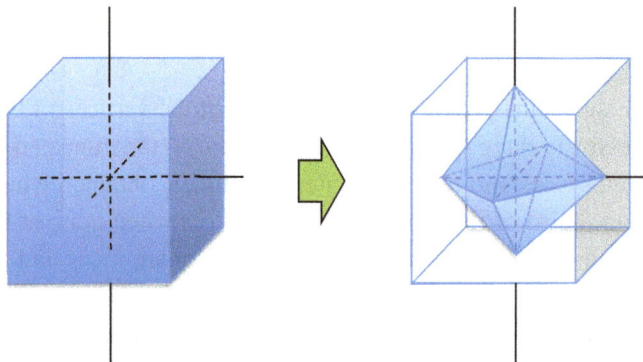

Fig. 1.2. Geometric construction of the octahedron, the dual form of the cube.

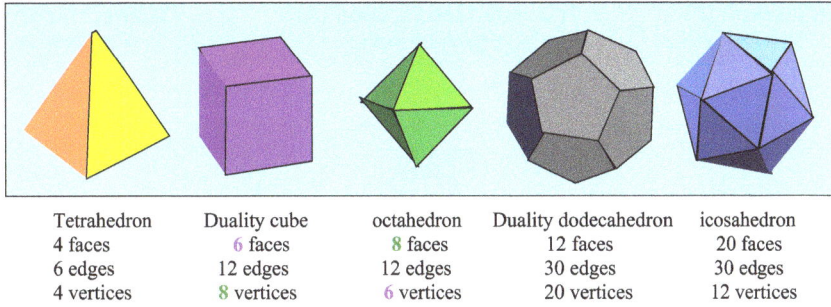

Tetrahedron	Duality cube	octahedron	Duality dodecahedron	icosahedron
4 faces	6 faces	8 faces	12 faces	20 faces
6 edges	12 edges	12 edges	30 edges	30 edges
4 vertices	8 vertices	6 vertices	20 vertices	12 vertices

Fig. 1.3. Duality relations and characteristics of the five Platonic polyhedra.

Using this rule, it is easy to show that the dodecahedron is the dual form of the icosahedron and *vice-versa*. On the contrary, the dual form of the tetrahedron is the tetrahedron itself (Fig. 1.3).

From Fig. 1.3, it is noteworthy that, for two dual forms, (i) the number of faces in one form becomes the number of vertices of the other and *vice-versa* and (ii) the number of edges remain invariant for the two forms. Whatever the polyhedra, the Descartes–Euler relation between the number F of faces, the number E of edges and the number V of vertices is always verified.

$$F + V = E + 2.$$

The above rules of truncation applied to the five Platonic polyhedra lead to the 13 Archimedian polyhedra (Fig. 1.4).

All the announced polyhedra are not represented here. Two are missing: the *snub cube* and the *snub dodecahedron*. Before truncation, a rotation of the faces is necessary. Figure 1.5 explains their construction, first in the case of the cube. After the surface burst of the cube (which is equivalent to an intermediary rhombicuboctahedron), the initial surfaces of the cube are rotated by 16°. In this case, the squares of the intermediary rhombicuboctahedron are transformed into two triangles with a common edge, and the snub cube is formed.

In Fig. 1.4, under the names of each polyhedron, a succession of numbers appears in red. They correspond to the *Schläfli notation* (Schläfli: Swiss mathematician (1814–1895)) which allows a reduced definition of these polyhedra, beside their complicated names. In a general way, the first numeral represents the type of face of each polyhedron (3 for triangle, 4 for square, 5 for pentagon, 6 for hexagon, etc...); the second gives the number of each type of face joining at a vertex. For instance, in the notation of a tetrahedron: [3,3], the first 3 means triangle and the second that three triangles meet at a vertex. For the truncated tetrahedron (noted $3^1 \cdot 6^2$ or (simpler) $3 \cdot 6^2$), 3 and 6 indicate that the faces are both triangular and hexagonal, and that there is one triangle and two hexagons (superscript

Tetrahedron (P)
4 triangles
3,3

Truncated tetrahedron (A)
4 triangles; 4 hexagons
3.6^2

Octahedron (P)
8 triangles
3.4

Dodecahedron (P)
12 pentagons **5,3**

Truncated Dodecahedron (A)
20 triangles; 12 decagons **3.10^2**

Icosidodecahedron (A)
20 triangles; 12 pentagons **$(3.5)^2$**

Cuboctahedron (A)
8 triangles; 6 squares **$(3.4)^2$**

Great rhombicuboctahedron (A)
6 octogons; 8 hexagons; 12 squares **4.6.8**

Small rhombicuboctahedron (A)
8 triangles; 18 squares **3.4^3**

Icosahedron (P)
20 triangles **3.5**

Truncated icosahedron (A)
12 pentagons; 20 hexagons **5.6^2**

Icosidodecahedron (A)
20 triangles; 12 pentagons **$(3.5)^2$**
(once more!)

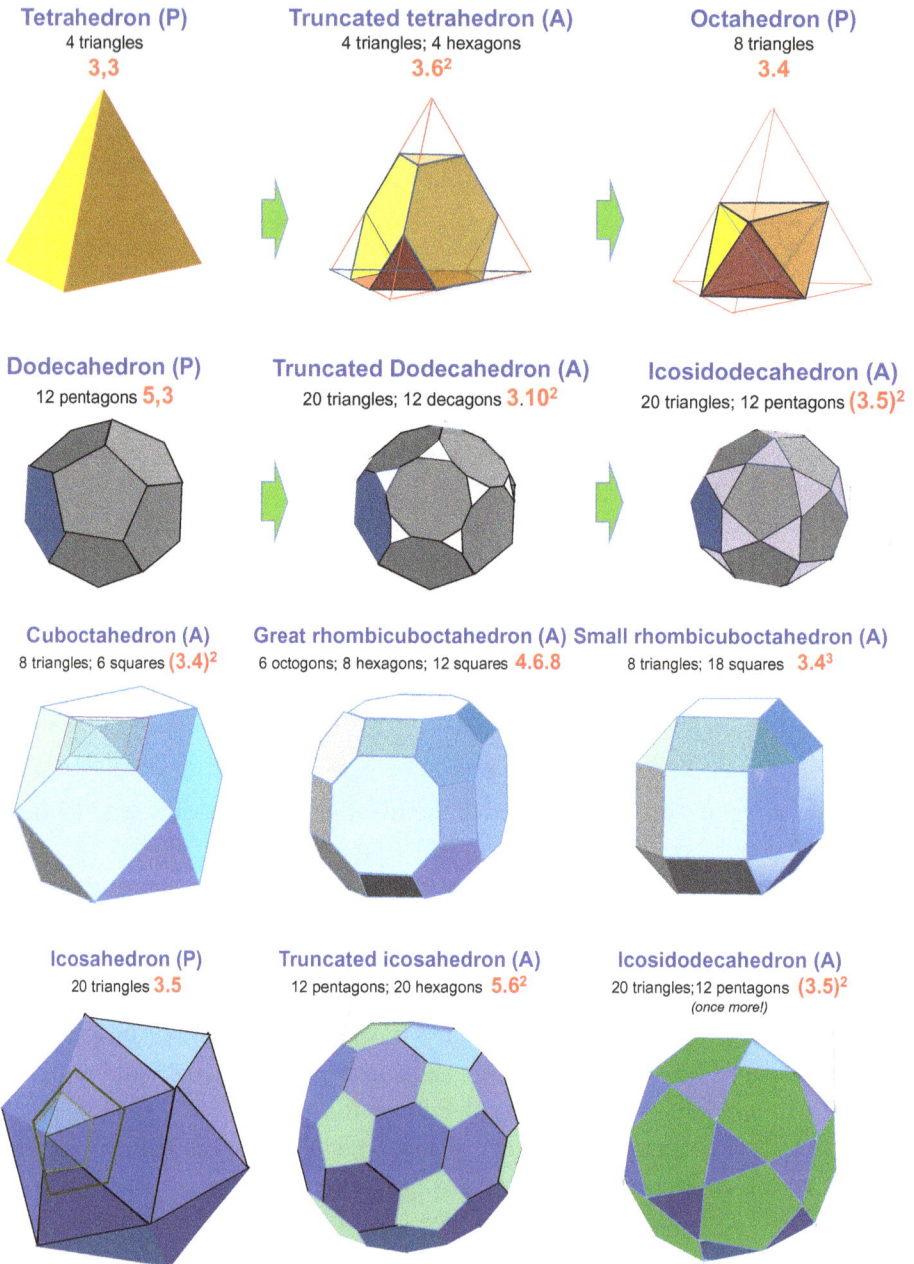

Fig. 1.4. Construction and names of the Archimedian polyhedra obtained by truncation of Platonic polyhedra.

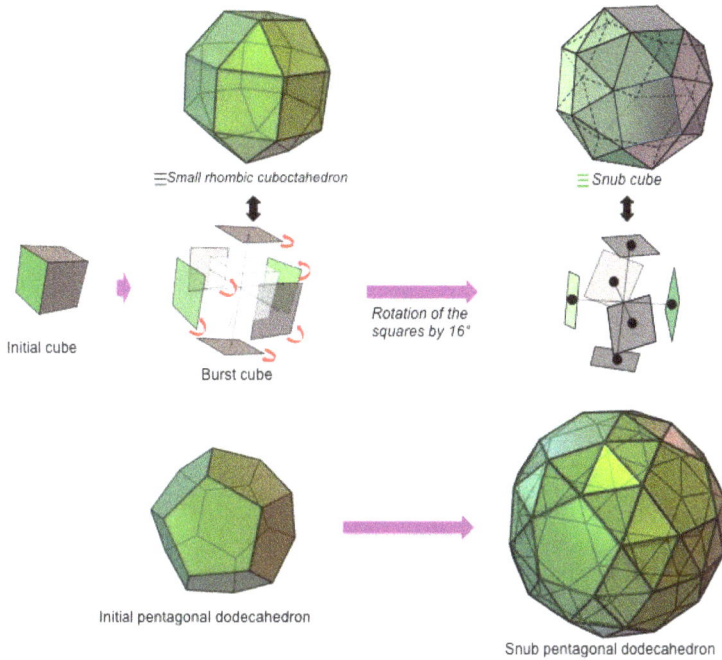

Fig. 1.5. Construction of the snub cube and of the snub dodecahedron by burst and rotation (the sign ≡ means here equivalent to).

numerals) joining at a vertex. The notation of the dual forms of the Archimedian polyhedra, which begins with the letter **V**, gives the number of edges intersecting at each type of vertex and the number of times when each type is associated to a face. For example, the Schläfli symbol of the rhombic dodecahedron (rhombic from the Greek *rhombus* (lozenge)), which is the dual form of the cuboctahedron, is V$(\mathbf{3.4})^2$. This means that **2** vertices are formed by the intersection of **3** edges, and that **2** other vertices correspond to that of **4** edges (Fig. 1.6).

Appendix 1 at the end of the book provides, for each polyhedron, its various geometrical characteristics and also their two-dimensional development for the

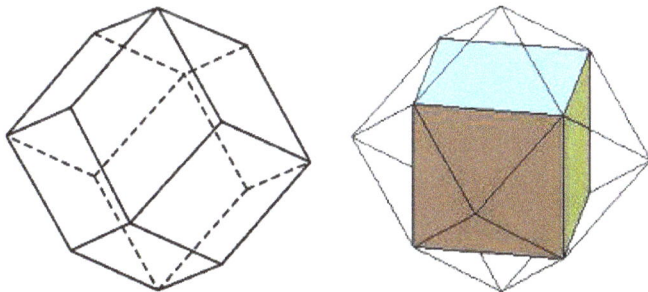

Fig. 1.6. The rhombic dodecahedron, dual form of the cuboctahedron. It can be generated from a cube by adding on each of its faces a hemi-octahedron whose edges are identical to those of the cube.

readers who have difficulties to see in three dimensions and need three-dimensional models which are easy to build with paper.

One could play for a long time with these mathematical games. Starting from these mathematical objects, it is now time to approach more chemistry.

1.2 From Polygons to Polyhedra for Some Chemistry

The above polyhedra are simply the three-dimensional assembly of regular polygons. In our day life, we are, even unconsciously, surrounded by objects having these polygonal shapes: a wire netting, the red tiles of a floor are hexagons, the juxtaposition of which creates the pavement (Fig. 1.7).

These shapes are also encountered in the organization of matter at the atomic level. In this sense, chemistry could be considered as a decoration of their geometry. Indeed, if spheres of increasing sizes are placed at the vertices of the hexagon, this gives an idea of the atomic arrangements. A hexagon decorated by interpenetrating CH groups represents the benzene molecule; an assembly of edge-shared hexagons decorated by interpenetrated carbon atoms is equivalent to one layer of the lamellar structure of graphite (Fig. 1.8). Note that, in this case, the center of each hexagon is empty because a supplementary carbon atom is too large for being placed at this center. It is the first similitude between geometry and chemistry. It is not fortuitous, and one can go further by stacking these sheets, which implies a passage from the plane to the three-dimensional space.

Fig. 1.7. Examples of hexagons in our daily life.

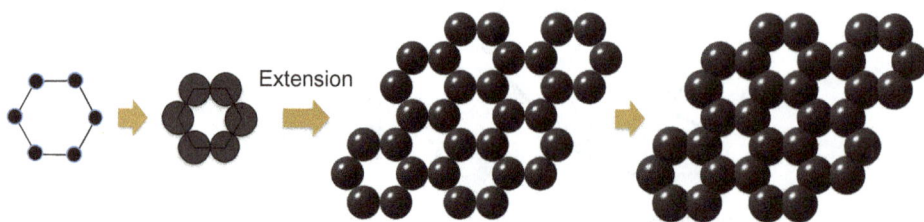

Fig. 1.8. Decoration of hexagons by spheres of increasing radii.

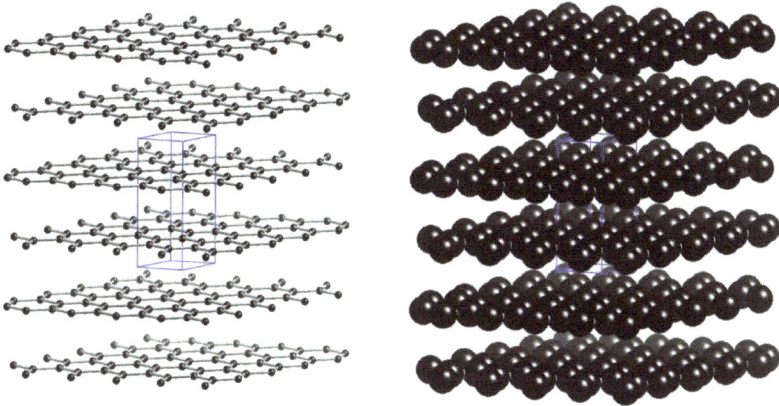

Fig. 1.9.   (Left) Balls and sticks and (right) space filling description of graphite.

There are two ways to do this stacking: The first aims at a packing as dense as possible. Each sphere of the top layer then comes in contact with the six which form the hexagon of the down sheet (Fig. 1.9). The resulting arrangement is close, but not exactly that of graphite. Indeed, in graphite, the sp^2 hybridization of carbon leads to two types of bonds: short σ bonds within the layer and longer π bonds between layers.

In the ideal case, spheres are strictly tangent. This implies that the space at the center of each hexagon becomes sufficient for hosting another identical sphere. In each plane, the original hexagonal net becomes a triangular one. In terms of stacking, each atom of the top layer projects on the center of one over two equilateral triangles of the down layer. It will be seen later, after the definition of a crystallographic cell with linear parameters a, b, c, that one passes from one plane to the other by a $\frac{1}{3}a + \frac{2}{3}b + \frac{1}{2}c$ translation. For the moment, the two successive planes are noted A and B and are defined a [A-B] stacking, which means that the third layer is identical to the first (Fig. 1.10).

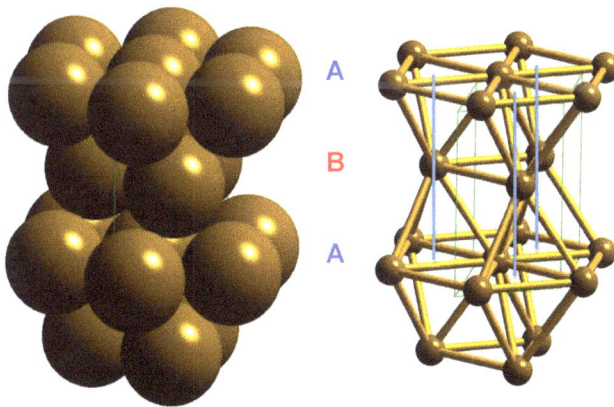

Fig. 1.10.   Dense hexagonal stacking of tangent identical spheres.

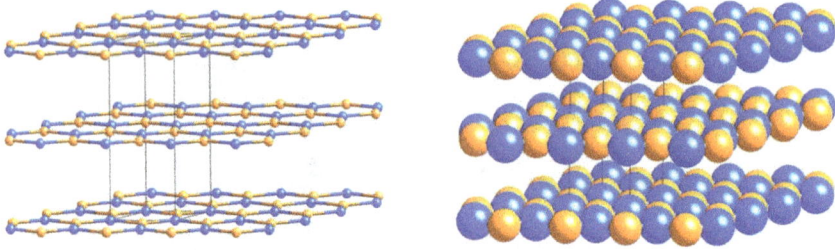

Fig. 1.11.   Noncompact hexagonal stacking of tangent spheres. The different colors correspond to boron and nitrogen atoms in boron nitride.

The second way (Fig. 1.11) aims at the strict superposition of two successive layers. In this case, their stacking is not compact since, from one layer to the other, one atom is tangent to only two spheres instead of six in the first case, one above and one below. This arrangement is encountered in the structure of boron nitride (BN).

As it was the case for hexagons, squares macroscopically surround daily life. The same concept of decoration leads to what is called a primitive cubic arrangement noted by P (Fig. 1.12).

The cube! A fascinating polyhedron! It is the dual form of the octahedron [Fig. 1.13(a)]. It can also contain the tetrahedron, but this time with no relation

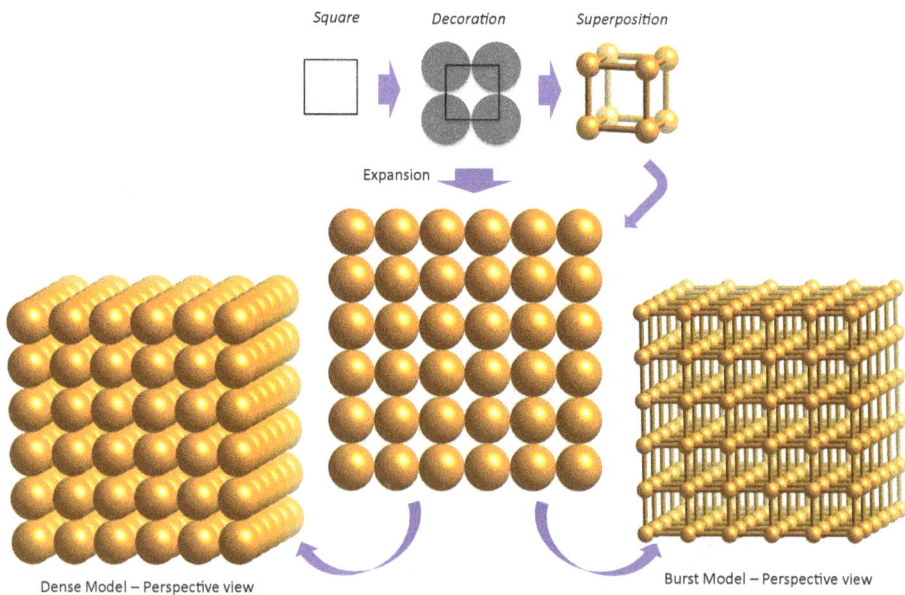

Fig. 1.12.   From the square to the primitive cubic arrangement by decoration, stacking and expansion.

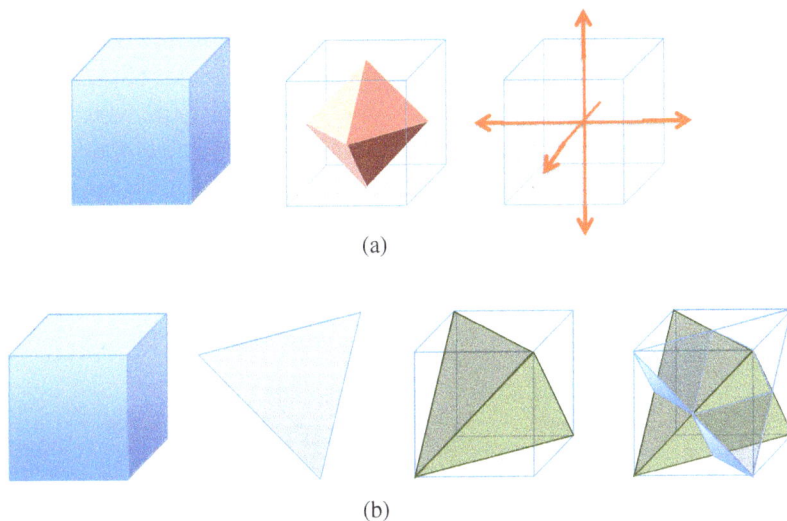

Fig. 1.13. (a) Cube-octahedron duality; (b) the two types of tetrahedra deduced from the cube.

of duality between the two forms. The tetrahedron is obtained by joining two opposite vertices of each face of the cube. Depending on the starting vertex, two orientations of the tetrahedron are possible [Fig. 1.13(b)].

Attempts of decoration can be extended to more complex objects, for instance, to regular octahedra for which spheres are envelopes. This means that what was true for spheres will be true for octahedra (Fig. 1.14). This assembly of octahedra is found in the structure of rhenium trioxide (ReO_3).

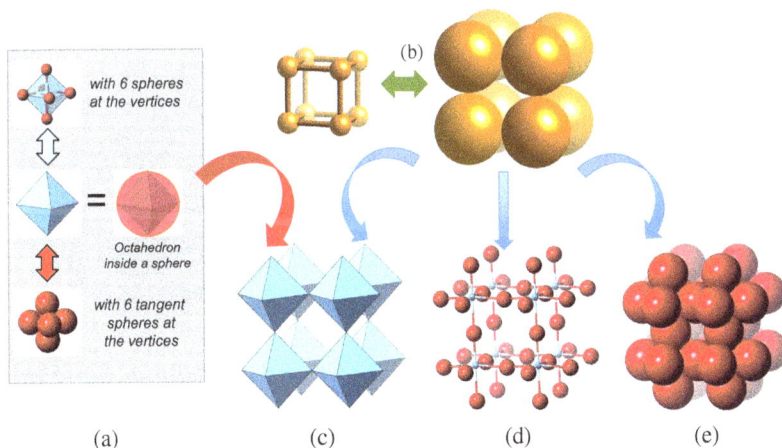

Fig. 1.14. (a) The different representations of an octahedron within a sphere; (b) starting from the primitive cubic arrangement, the substitution of spheres by octahedra leads to the arrangement in terms of polyhedra (c) of balls and bonds (d) and in terms of tangent spheres (e).

In this type of structure, the stacking is of course not compact. If one notes a, b and c to be the lengths of the edges of the original cube, the compacity can be increased if one atom of the upper plane projects on the center of the square of the lower plane. This corresponds to a $\frac{1}{2}a + \frac{1}{2}b + \frac{1}{2}c$ translation between two layers. In these conditions, each atom is in contact with eight neighbors: four atoms above it and four atoms below (Fig. 1.15). The resulting arrangement is called **centered cubic** (symbol cc) because one atom of the second layer is at the center of its eight tangent neighbors.

This disposition of atoms is encountered in the structures of a large number of metals, in particular, in the α low temperature structure of metallic iron (α-Fe).

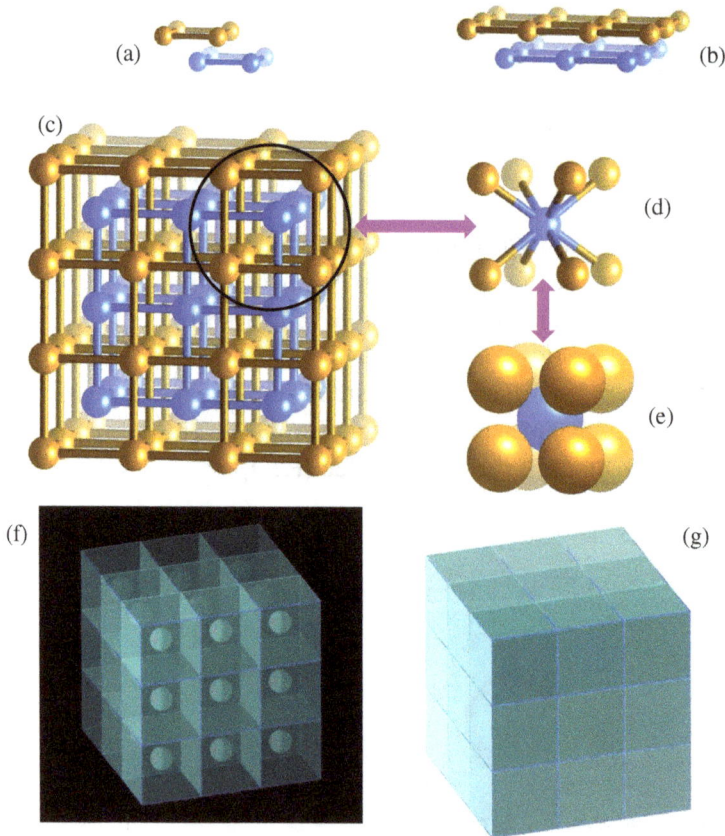

Fig. 1.15. (a,b) Dense stacking obtained by shift of two consecutive squares or layers of squares. Different colors are used for evidencing the shifting of two adjacent sheets of *identical* atoms; (c) extended view of the arrangement from which one cube was isolated (circle), represented both in spheres and bonds (d) and in terms of tangent spheres (e). Figures (f) and (g) represent two polyhedral views of the arrangement: (f) translucent and (g) filled polyhedra. This polyhedral representations of primitive cubic and centered cubic arrangements allow their descriptions in the three *dimensions* of the space as the aggregation by sharing faces of cubes, empty in primitive cubic, and filled in the other case.

Remark: If two consecutive layers are built from the atoms of different chemical nature, the arrangement is no longer centered cubic, but becomes the interpenetration of two primitive cubic lattices. This is the case of cesium chloride CsCl, in which Cesium occupies the center of a cube with Cl⁻ anions at its vertices [Figs. 1.15(f) and 1.15(g)].

Like children playing games with their cubes, crystal chemists can also play with these atomic cubes. Imagine that, starting from the assembly of cubes of Fig. 1.15(f), one eliminates regularly in the three dimensions of the space one filled cube over two.... Chemically, one of the most frequently encountered structures of the inorganic world is obtained: the fluorite structure. Its aristotype (idealized drawing representing a large family of equivalent structures) is calcium fluoride (CaF_2).

This frequent structural type is found, either perfect (aristotype) or distorted (ettotype), in very different chemical families (fluorides, oxides, cyanides, nitrides, carbon-oxynitrides). Figure 1.17 provides some illustrations.

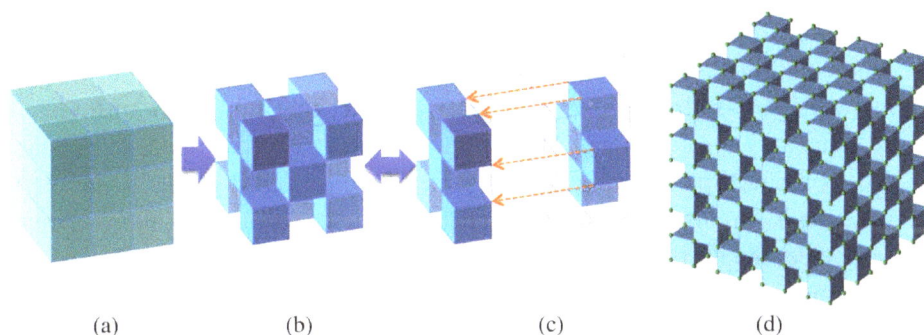

(a)   (b)   (c)   (d)

Fig. 1.16. Polyhedral description of the relation between the structures of (a) CsCl and (b,d) fluorite CaF_2. (c) The details of the connection by vertices and edges between two successive layers.

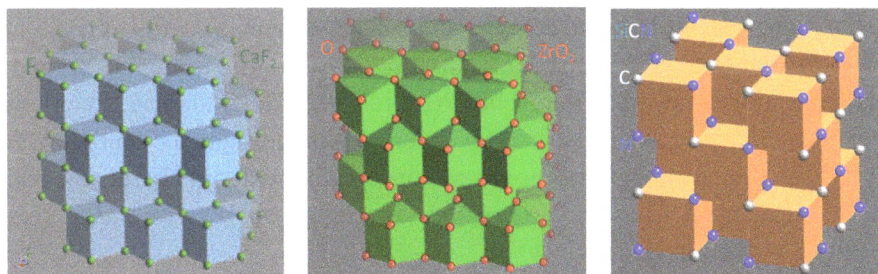

Fig. 1.17. Some examples of chemical compounds exhibiting the fluorite structure: (left) CaF_2, (middle) distorted ZrO_2 and (right) carbon nitride SiCN. In each case, the metal occupies the center of the cube.

Fluorite structure CaF$_2$ Primitive cubic net of F$^-$ Face centered cubic net of Ca^{2+}

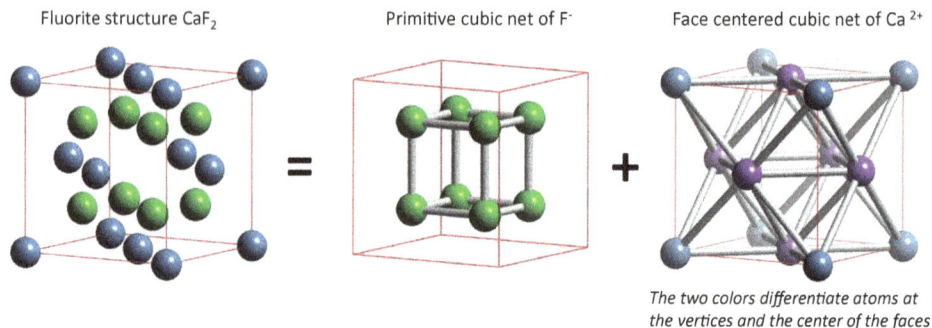

The two colors differentiate atoms at
the vertices and the center of the faces

Fig. 1.18.   Decomposition of the fluorite network into its two subnetworks: the primitive cubic F$^-$
and face centered cubic Ca^{2+}.

A first lesson from that: it is not only the chemical nature of atoms which determines the structures, but other factors (geometrical, electronic...) which govern the arrangements, as it will be explained later in this book.

The cubic fluorite structure, described from atoms and bonds can also be considered in terms of two overlapped subnetworks (Fig. 1.18): the cationic Ca^{2+} (in blue), and the anionic F$^-$ (green). When examined separately, if the fluoride subnetwork is the already known primitive cubic, that of Ca^{2+} is different. It corresponds to a cube with cations on both the vertices and on the center of each face. It is a new type of arrangement, named **face centered cubic**. Its acronym is *fcc*. Such an arrangement is found in the structures of many metals. For instance, the high temperature structure of iron metal, labeled γ-Fe, adopts this topology whereas the low temperature form α-Fe was centered cubic (see above).

Another way of looking at the *fcc* topology is to consider it in terms of dense packing in the direction of the diagonal of the cube. Three different (by their relative positions) layers appear (Fig. 1.19). They are noted *A*, *B* and *C*.

One can also apply the concept of decoration to this structure type by replacing the single spheres by more complex units. One of them is the fullerene C$_{60}$, the new molecular form of carbon (Fig. 1.20) discovered by H. Kroto, R. Curl and R. Smalley who were awarded the Nobel Prize in 1992. This fullerene is composed of the assembly of groups, each of 60 spheres. It is the same *fcc* topology. This remarkable homothety gave rise to a new concept: the **scale-up chemistry**, which will be detailed in Chapter 7, once the knowledge on supplementary crystal chemistry is developed.

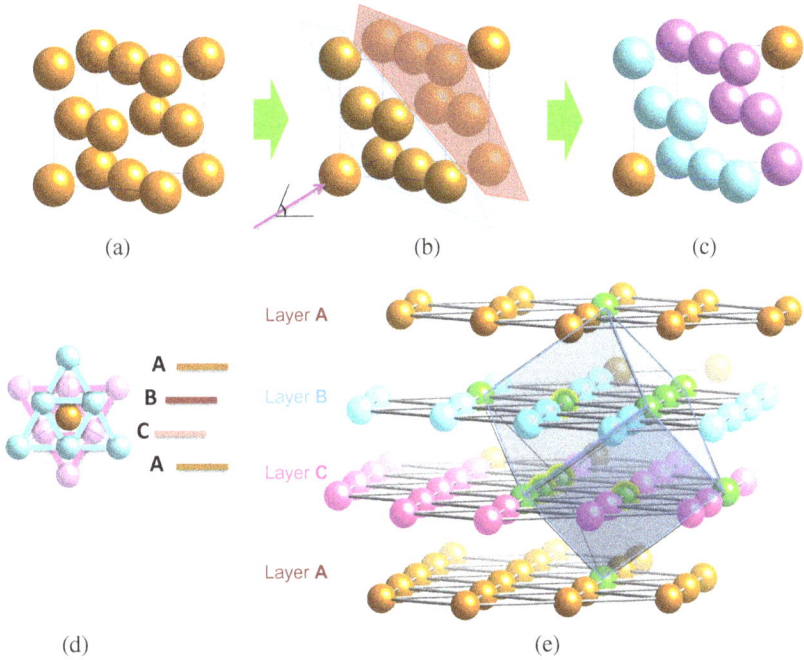

Layer A

Layer B

Layer C

Layer A

(a)   (b)   (c)

A
B
C
A

(d)   (e)

Fig. 1.19. (a) Perspective view of the face centered cubic structure; (b) evidence of the compact triangular layers with different colors in fcc; (c) perspective view using different colors for each layer; (d) the projection along the diagonal of the cube; (e) the fcc structure defined from the succession of layers *A*, *B* and *C*. The vertices of the cube correspond to green spheres.

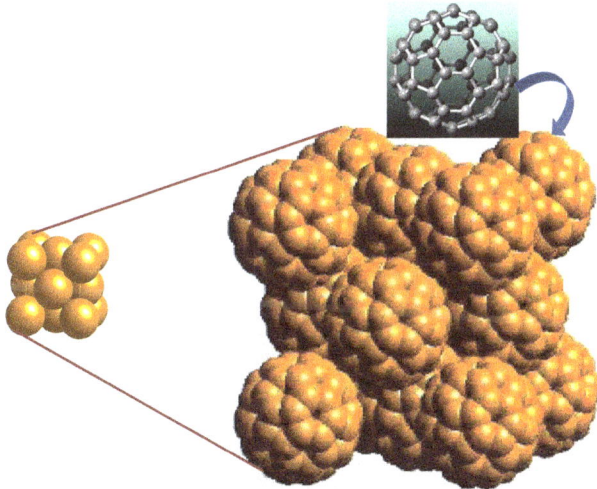

Fig. 1.20. Illustration of the homothety existing between the structures of γ-Fe and fullerene C_{60}, the topology remaining invariant. The insert on top right corresponds to the balls and stick representation of fullerene.

By the way, the examples of α-Fe and γ-Fe help to suppress a common idea, usually thought of by young chemists. A given chemical compound can exist in different crystallographic forms, depending on the conditions of preparation. The terms **allotropic varieties or polytypes** are used for defining this phenomenon. One illustrative example concerns carbon, illustrated as Paving 2 at the end of this chapter.

The first part of this chapter is now completed. This allows a qualitative familiarization with major geometric shapes and a progressive introduction to dense or quasi-dense packings in relation with topology, which means the analysis of shapes, their mutual relations, in particular, those concerning Platonic and Archimedian polyhedra. It is time now to go further and to characterize them more quantitatively. This implies some further definitions.

1.3 From Qualitative to Quantitative. Some Definitions

First, periodicity and its related vocabulary.

Look at a Persian carpet (Fig. 1.21)! Its central part is limited by a series of squares, the vertices of which are occupied by circles. Inside each square, on a blue background, appear stylized flowers with five petals. They are all identical, whatever the direction, from left to right or from the bottom to the top. The carpet therefore corresponds to the regular juxtaposition of squares in two directions. This juxtaposition is named **translation**, and the building unit (here the squares) is called the **motif**.

What is true in two dimensions for the carpet at the macroscopic scale is also true at the atomic level in three dimensions. Take a sphere, equivalent to one atom, copy it in the three *dimensions* of space and render them tangent (Fig. 1.22). The infinite repetition of this operation generates a macroscopic assembly of atoms (cubic in Fig. 1.22) …

Fig. 1.21.   A Persian carpet, an example of two-dimensional periodicity.

Fig. 1.22. Juxtaposition by translation of a spherical motif in the three *directions* of space generating a macroscopic solid.

In this operation, the sphere is the motif and the distance between two consecutive spheres determines the translation parameter in each direction. The volume defined by the three vectors of translation and the angles between them is called **the cell**. Its orientations and its labels obey international rules (Fig. 1.23).

The cell is therefore the minimum volume which, by translation, describes the macroscopic solid. From arguments of symmetry, only seven types are possible in three dimensions. They are called the **seven crystalline systems**. They are illustrated with their characteristics in Fig. 1.24.

Incidently, what is the name of the scientist at the origin of the concept of periodicity at the atomic scale? He was a French mineralogist, René-Just Haüy, in 1774! Some historical details on science belong to it!

Whereas his younger brother Valentin (1745–1822) dedicated his life to blind people, René-Just, passionate about mineralogy, thought that it was possible to classify the crystals of minerals as a function of their external shape. He had the idea to take a very nice rhombohedral crystal of calcite (a form of calcium carbonate $CaCO_3$) and to break it with a hammer. He observed that the resulting smaller crystals had exactly the same shape as the original crystal. By extrapolation, he suggested that, after an infinite repetition of this operation, the organization

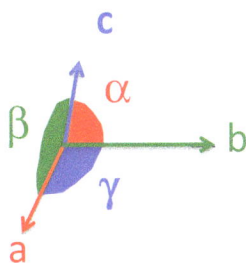

Fig. 1.23. International convention for the representation of a crystallographic cell. The angle α is opposite to *a* axis, β is opposite to *b* axis and γ is opposite to *c* axis.

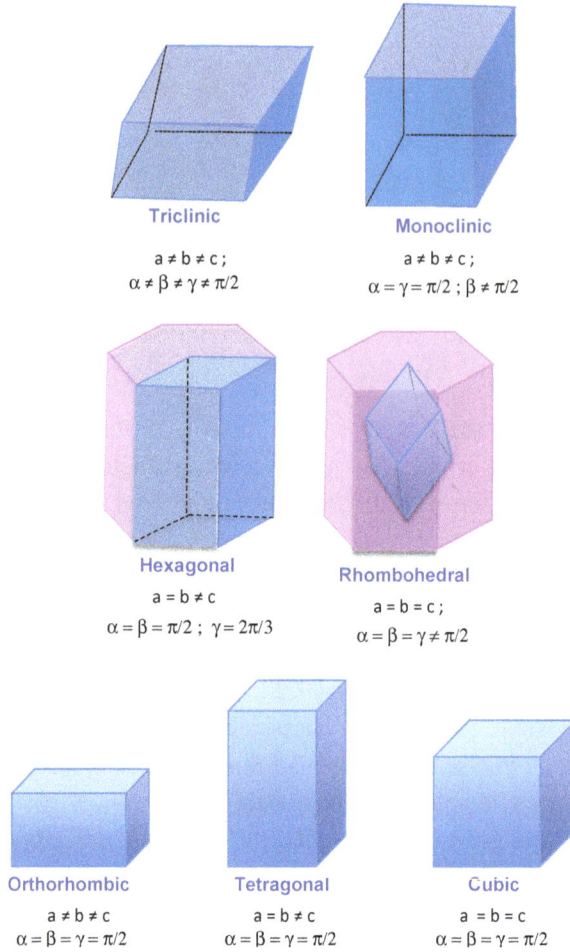

Fig. 1.24. The seven crystalline systems. The hexagonal cell can be reduced to a lozenge-based prism; the rhombohedral cell can be described in a multiple hexagonal cell. The orthorhombic and tetragonal cells are rectangle- and square-based prisms, respectively.

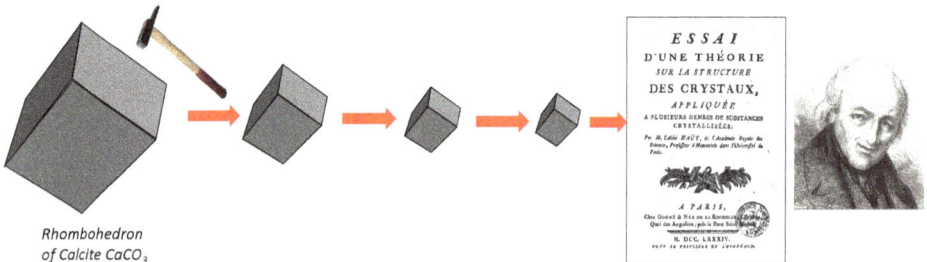

Fig. 1.25. The experiment of René-Just Haüy (1743–1822).

remained the same, without knowing what organization *meant. One must keep in mind that, at that time, the physical notion of atoms did not exist... It was only philosophers who spoke about atom, after Plato! It is thanks to Haüy and his contemporary Jean-Baptiste Romé de l'Isle (1736–1790), who had measured the angles between two faces of a crystal and observed their invariance whatever the relative development of these faces (Romé de l'Isle law) that geometric crystallography was born!...*

A more precise definition of **motif**?

It is every assembly, real or virtual, contained within the cell which, by translation in the three dimensions of space, regenerates the macroscopic solid. When real, the motif is a molecule, and the distances between molecules are longer than those existing within this molecule. One speaks of molecular motifs. When the motif is virtual, it cannot be distinguished inside the assembly (Fig. 1.26).

In face centered cubic, for instance, the motif is built up from four atoms: one at the origin and the three occupying the center of the faces containing the origin. They form a tetrahedron of atoms (Fig. 1.27).

When several types of atoms, associated to a chemical formula, exist within the cell, the minimal number of atoms which correspond to the formula is called **formula unit**, whatever its nature (real or virtual).

Keeping in mind all these definitions is essential for understanding the following. The corresponding names will now be used without recalling their significance.

It is now necessary to come back to quantitative considerations characterizing dense or quasi-dense packings.

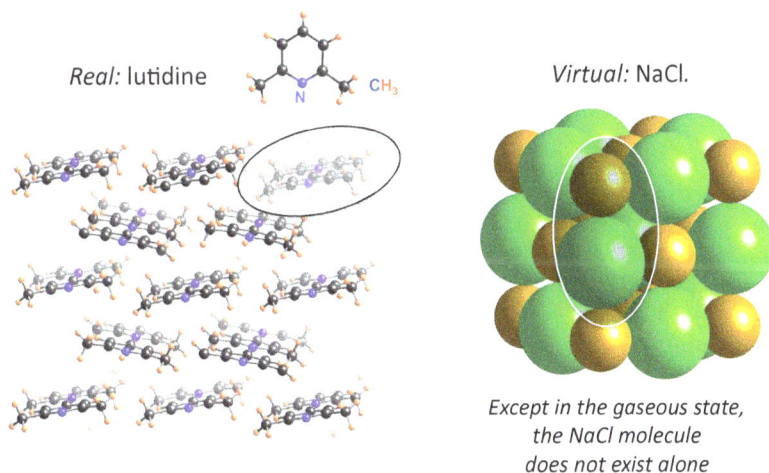

Real: lutidine   CH₃

Virtual: NaCl.

Except in the gaseous state, the NaCl molecule does not exist alone

Fig. 1.26. Examples of a real motif (lutidine) and a virtual one (NaCl). In the latter, a special NaCl group cannot be distinguished among others.

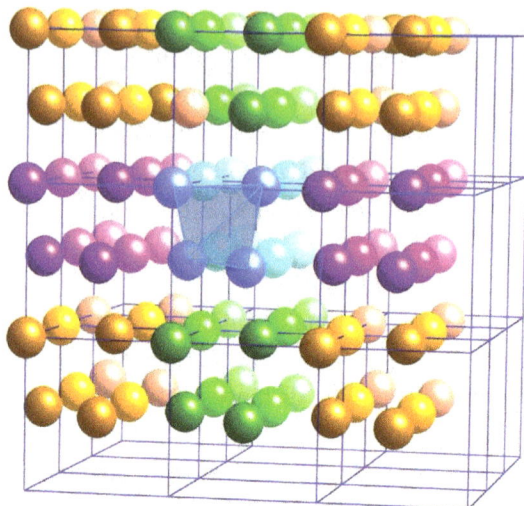

Fig. 1.27. Visualization, helped by different colors, of the motif and its translations generating the macroscopic fcc structure.

1.4 Simple Calculations on Dense Packings of Spheres

1.4.1 *Primitive cubic packing*

Consider first that primitive cubic packing is built up from tangent identical spheres of radius R. In this packing, the translation vector a is the same in the three dimensions of space. Its value is $2R$, and corresponds to the length of the edge of the cube. This length is named the **parameter** of the cubic crystallographic cell.

In the latter, the space is not fully occupied. There is a void at the center of the cell. It is called **interstice** (Fig. 1.28).

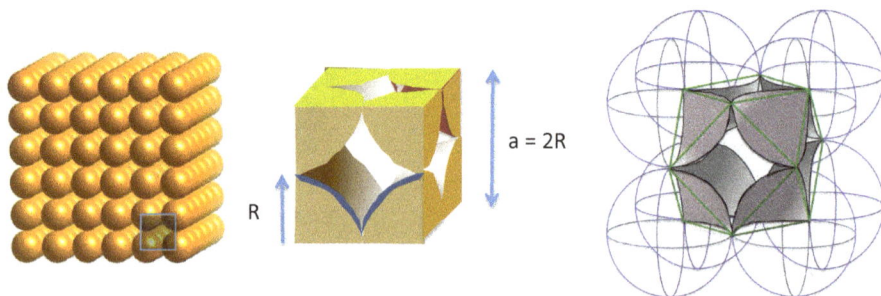

Fig. 1.28. (Left) Extended primitive cubic stacking, with the cell represented in grey; (center and right) illustration of the real shape of the interstice, which is a concave cuboctahedron.

Only fractions of the spheres at the vertices of the cubic assembly (here 1/8) are effectively inside the cell. As there are eight vertices for the cell, this is equivalent to one sphere (with volume $V = \frac{4}{3}\pi R^3$) inside the cell, the volume of the latter being $(2R)^3$. The unoccupied volume is therefore:

$$(8R^3 - 4/3\,\pi R^3) = 4R^3(2 - \pi/3) = 4R^3(0.9528).$$

The fraction of vacuum is then $[4R^3 \times (0.9528)/8R^3]$ or 47.64%. The primitive cubic arrangement is not really dense! But, can the interstice be filled?

In a primitive cubic, the center of the interstice is at the center of the cell. Its filling requires to find the maximum radius $\mathbf{r}$ of an atom which could occupy the void without perturbating the primitive arrangement. Considering that the spheres, with their $\mathbf{R}$ radius, are tangent along an edge of the cube, the inserted atom, with its $\mathbf{r}$ radius, must be tangent to all the initial spheres. This means that the length of the diagonal of the cube (edge: $a = 2R$) will correspond to $2R + 2r = a\sqrt{3} = 2R\sqrt{3}$. Therefore:

$$\mathbf{r/R} = {}^{(2\sqrt{3})} - \mathbf{1} \quad \text{or} \quad \mathbf{r = 0.732R}.$$

If $\mathbf{r}$ is larger than this value, the spheres with radius R will no longer be in contact. On the contrary ($r < 0.732R$), the inserted sphere will have some degrees of freedom within the interstice.

1.4.2 *The centered cubic packing*

In this case, the spheres are no longer in contact along the edges like in primitive cubic, but this time the contact occurs along the diagonal of the cube. Therefore, here, $4R = a\sqrt{3}$. The centers of the interstices are now at the center of the faces. After an extension to neighboring cells, it becomes clear that six spheres surround the empty space, forming a **flattened regular octahedron** (Fig. 1.29).

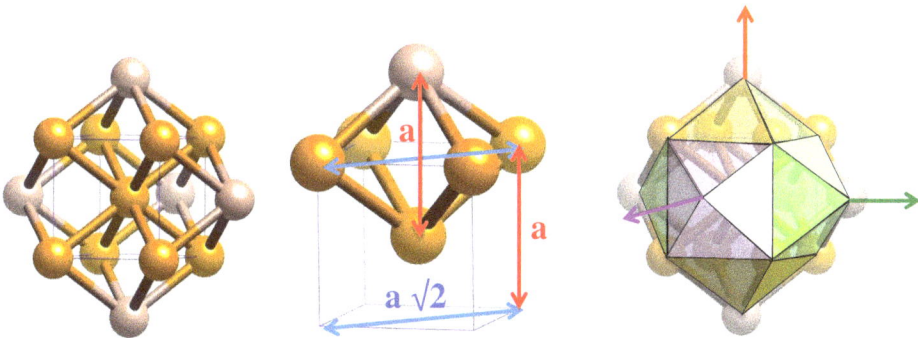

Fig. 1.29. Different views of the cubic centered arrangement and the polyhedral environment of the interstices.

What is the percentage of compacity for the cell? Within it, besides the $8 \times 1/8$ sphere which correspond to the atoms residing at the vertices of the cube, there is the central atom in its integrality. This time, two atoms are effectively inside the cell. This represents the motif of *cc*. The volume of matter is $8/3 \cdot \pi R^3$, within a cell of volume $64R^3/3\sqrt{3}$ (the cell parameter *a* being $4R/\sqrt{3}$). The percentage of matter is then $\pi\sqrt{3}/8 = 0.6801$, which represents almost 32% of void in *cc* instead of the 47. 64% in primitive cubic.

Once placed at the center of the flattened octahedron, the inserted sphere, with its **r** radius, will be tangent to only the closer two separated by distance *a*. Therefore, $2R + 2r = a = 4R/\sqrt{3}$ and

$$\mathbf{r/R} = \left(2/\sqrt{3}\right) - 1 \quad \textbf{or} \quad \mathbf{r = 0.155R}$$

which is extremely small! This will explain why only a very small percentage of carbon can be introduced in the metallic structure of iron α, which corresponds to the formula of steel.

1.4.3 *The face centered cubic packing*

In this arrangement (Fig. 1.30), the atoms (radius *R*) are tangent along the diagonals of the faces of the cube. Consequently, $4R = a\sqrt{2}$ (or $a = 4R/\sqrt{2}$ and $V = 64R^3/2\sqrt{2}$). In terms of number of atoms effectively inside the cell, besides the $8 \times 1/8$ sphere corresponding to the atoms at the vertices of the cube, there are also six atoms on the faces. For them, only one-half of their spheres resides in the cell, which means that three supplementary atoms must be added to the first one. In other words, within the cell, there are effectively four atoms. The volume of matter is then $16/3 \cdot \pi R^3$ in the cell volume of $64R^3/2\sqrt{2}$. After simple calculations, this gives a ratio of matter of $\pi\sqrt{2}/6 = 0.7401$, and therefore 26% of void. **It represents the densest cubic packing** and one of the two denser packings when hexagonal packing is also considered (see below).

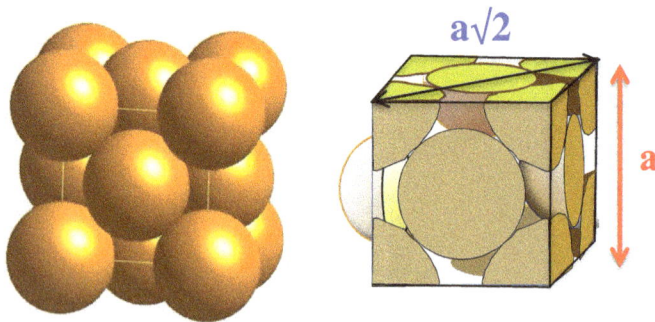

Fig. 1.30.   Different views of *fcc* packing and its regular octahedron interstice.

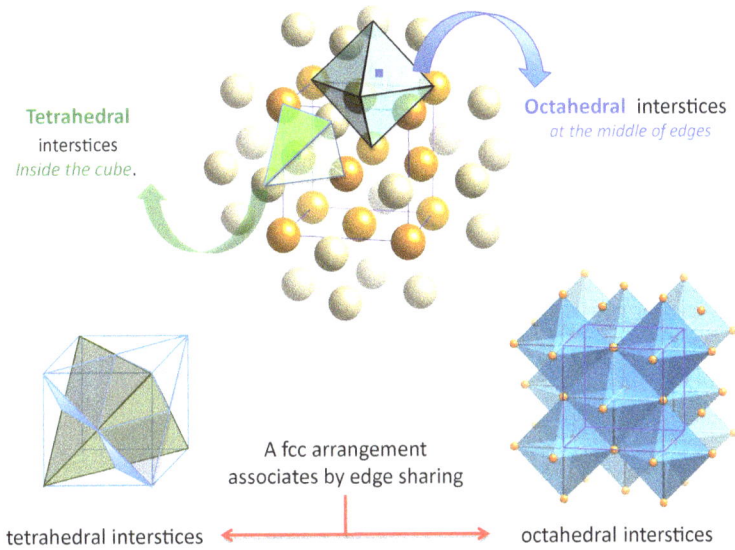

Fig. 1.31. Localization of the positions of the tetrahedral and octahedral interstices in the *fcc* structure with the description of these arrangements in terms of succession of interstices of the same type.

However, compared to the primitive and centered cubic networks, the situation is more complicated with *fcc* frameworks. There are this time, two types of interstices: one is, of course, octahedral, but there is also another one, which is tetrahedral (Fig. 1.31).

A calculation is now needed for evaluating the radius r of the atoms that can be introduced in the center of each type of interstice. For the octahedron, it is obvious since one edge corresponds to $2R + 2r$, the spheres being tangent along a diagonal of face $(4R = a\sqrt{2})$.

$$r/R = \sqrt{2} - 1 = 0.414.$$

For the tetrahedron:

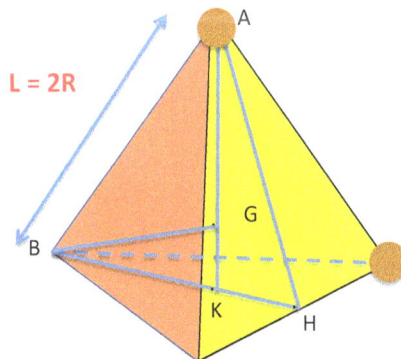

The barycenter G of the four tangent spheres (radius R) is situated at ¼ of AK starting from K. From the relations existing for equilateral triangles, one can say:

$$AH = L\sqrt{3}/2 = R\sqrt{3}, \quad KH = BH/3 = AH/3 = (R/3).\sqrt{3} \quad \text{and} \quad BK = (2R/3).\sqrt{3}.$$

As *Pythagoras theorem* says $AK^2 = AH^2 - KH^2 = 9\,KH^2 - KH^2 = 8KH^2$, then

$$AK = KH\sqrt{8} = R\sqrt{(8}/3$$

and $GK = AK/4 = R[\sqrt{(8}/3\,]/4$, otherwise $[GK^2 = R^2/6]$.

As the value of $BG(r + R)$ is also $BG^2 = GK^2 + BK^2$, therefore

$$BG^2 = GK^2 + BK^2 = R^2/6 + 4R^2/3 = 9R^2/6 \quad \text{or} \quad BG = 3R/\sqrt{6}.$$

In other words, $BG = r + R = 3R/\sqrt{6}$ then **r/R = 0.224**.

When the spheres with radius R remain tangent, the tetrahedral interstice is very small compared to its octahedral homologue and can be occupied only by atoms of small radii.

1.4.4 *The hexagonal packing*

This time, the cell is not a cube. It could be an hexagonal prism but, as already stated, it can be reduced to a lozenge-based prism with an edge $a = 2R$ (and surface $4R^2\sqrt{3}/2$) and height c (Fig. 1.32). This height is twice that of the AK height of the tetrahedron seen above ($2R\sqrt{(8}/3$. The volume of this lozenge-based prism is therefore $8R^3\sqrt{2}$. Moreover, the ratio $\sqrt{(8}/3 = 1.633$ (often called c/a) between the height c of the cell and the parameter a, length of the edge of the base lozenge, characterizes a perfect hexagonal packing.

By application of the calculations as above, it is easy to show that the cell contains two atoms: one coming from the contribution of vertices, the other being inside the cell. It is also easy to show that the percentage of matter is once more $(\pi/6).\sqrt{2} = 0.7401$, as already found for *fcc*. **fcc and hc packings have identical compacities**. This result is not really surprising because these two structure types can be described by the stacking of the same dense triangular layers, but arranged in two different ways.

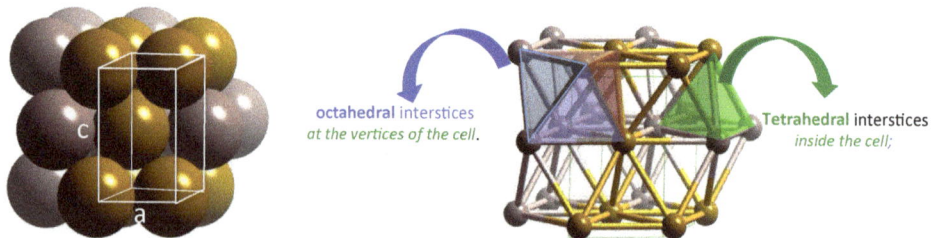

octahedral interstices
at the vertices of the cell.

Tetrahedral interstices
inside the cell;

Fig. 1.32.   Perspective view of the hexagonal dense packing and localization of its two types of interstices.

This identity implies to find in the hexagonal packing the same two types of interstices, tetra- and octahedral, as those found in *fcc*. This will lead also to the same values of **r/R** (0.224 and 0.414) for the tetra- and octahedral interstices.

1.5. Some Remarks

1.5.1 *Parameters influencing the dimension of the particles*

Since the beginning of this chapter, dense packings were described in terms of spheres that were often assimilated to atoms for approaching the chemical nature of the solid matter. However, when spheres were used, no dimensions were associated with them. If atoms (or ions) are considered, one exits the theory and enter reality. This implies that their dimensions, in the vicinity of the angström ($1 \text{ Å} = 10^{-10} \text{ m} = 0.1$ nm) must be taken into account. Indeed, for the same element, their size depends both on their physical nature (cation, neutral element, anion) and also on their neighborhood. The number of nearest neighbors (named coordinence or **coordination number** with CN as acronym) affects the size of the particles. It will be shown later that the polyhedra associated to this number are often regular polyhedra.

The dimension of the particles found in the solid strongly depends on their physical state in the resulting associations. Indeed, compared to the neutral element, the cation, which has lost some peripherical electrons, will have a smaller size. It is the contrary for the anion (Table 1.1). This trivial remark underlines a fact that is not taken enough into consideration: the geometrical aspects observed in the structures will strongly depend on the nature of the chemical bonds in which the particles are engaged.

This table clearly shows that the size of cations and anions increases when the coordination number increases. These values induce a relation between the sum of the theoretical sizes of two linked particles compared to the observed distances, and the nature of the chemical bond (ionic, covalent, van der Waals…). These various types of bonding correspond to various distributions of the electronic density around the particles involved in the bond (Fig. 1.33). This will sometimes allow to predict, at least qualitatively, some physical properties of the solid.

1.5.2 *Non-ideal dense packings*

The dense packings previously described do not represent the only cases encountered in the study of solids. They are only ideal examples, but, in the real solid, these packings offer many variations derived from the ideal cubic and hexagonal arrangements whose sequences were respectively noted *A*-*B*-*C* and *A*-*B* or *c* and *h*. Nature has a lot of imagination and exhibits many *fantasies*!

Table 1.1.   Values of the radius of some particles, as a function of their physical state and their coordination. [Values extracted from R. Shannon, *Acta Crystallogr.* 1976, vol. A32, p. 751].

	Metal	CN	Radius (Å)	Cation	CN	radius (Å)	Anion	CN	radius (Å)
Na	Na^0	12	1.91	Na^+	4	1.13	—	—	—
					6	1.16			
Al	Al^0	12	1.60	Al^{3+}	4	0.53	—	—	—
					5	0.62			
					6	0.675			
Fe	Fe^0	12	1.27	Fe^{2+}	4	0.77 HS	—	—	—
					6	0.92 HS			
				Fe^{3+}	4	0.63 HS	—	—	—
					5	0.72 HS			
					6	0.785 HS			
O	O^0	12	0.89				O^{2-}	4	1.24
								6	1.26
F	F^0	12	0.87				F^-	4	1.19
S	S^0	12	1.27				S^{2-}	6	1.70
I	I^0	12	1.60				I^-	6	2.06

HS: high spin; BS: low spin: notions of electronic configuration of ions, related to their optical and magnetic properties.

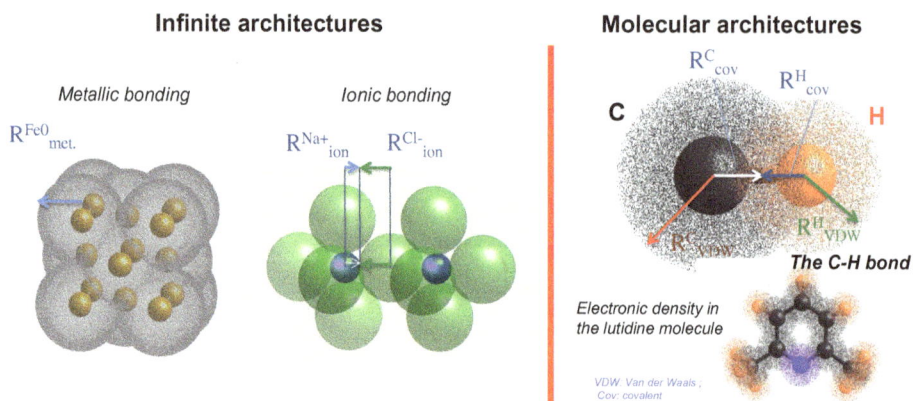

Infinite architectures

Metallic bonding

$R^{Fe0}_{met.}$

Ionic bonding

R^{Na+}_{ion} R^{Cl-}_{ion}

Molecular architectures

R^C_{cov} R^H_{cov}

C

H

R^H_{VDW}

The C-H bond

R^C_{VDW}

Electronic density in
the lutidine molecule

VDW: Van der Waals ;
Cov: covalent

Fig. 1.33.   Illustration of the different types of radii as a function of the type of chemical bonding in which they are engaged.

For instance, the sequence *A-B-A-C-* is observed in some niobate structures. It is called *double hexagonal* since it corresponds to the stackings of two simple hexagonal packings (*A-B-* and *A-C-*), but in which the positions of the *B* and *C* layers are different, leading to a doubling of the periodicity along the *c* axis. It is only one example among many others, but what was called *fantasies* a few lines above have in fact scientific reasons. **Never forget that the solid, whatever its complexity, must be thought thermodynamically in terms of energy**. Its most usual manifestation is, in terms of enthalpy, the lattice energy of a solid. As a function of the thermodynamic conditions to which it is submitted (temperature, pressure), the observed solid always corresponds to a minimum of free energy, where enthalpy and entropy play a role... The structural *fantasy* observed by the crystal chemist is in reality the qualitative expression of a rigourous calculation of an expert in thermodynamics...

Also, the values of the theoretical **r/R** ratios of ideal dense packings must not be applied strictly. They just represent an optimal value; for the thermodynamic reasons explained above, a tolerance of approximately 10% is accepted. If these ratios take slightly larger values, this will mean that the atoms are no longer tangent, while keeping their global topology.

1.5.3 *Real solid versus ideal solid*

Up to now, the ideal solid was considered. However, the young crystal chemist must always keep in mind that his studies will concern the real solid, even if its description tries to compare it to the ideal case, in the interest of simplification. Perfection rarely exists in the solid state! Coordination polyhedra are not always regular, the distances between atoms are not always equal... Despite that, the solid exists! It is its study which is of interest...

1.5.4 *Movements of particles*

Another established idea comes from a naive definition of the solid, synonym of immobility. It is not true! Indeed, except at 0 K, atoms in the solid vibrate around their equilibrium position. It is called **thermal motion**, which increases when the temperature increases. In the majority of the cases, the magnitude of the vibration is less than 1 Å. It can be isotropic or anisotropic. It is the first case of movement.

A second one, more important, occurs when the solid, submitted to the variation of either temperature or pressure, changes its structure. This movement is called **crystallographic phase transition**. It exists according to two types: the *displacive* phase transition, during which there is not any bond breaking in the structure; the movements, of weak magnitude (≤ 1 Å), usually concern rotations of the polyhedra (Fig. 1.34). In the second type, the *reconstructive* phase transition, some bonds are broken and lead to a different arrangement of the atoms.

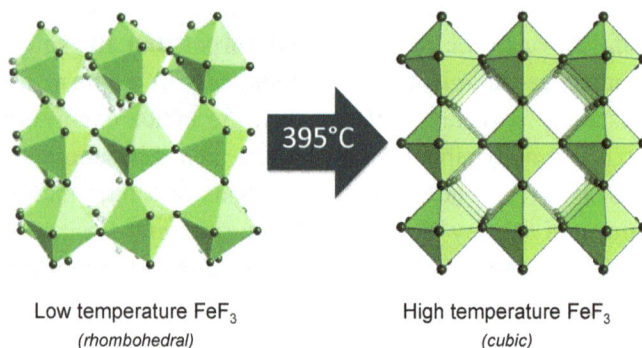

Low temperature FeF$_3$
(rhombohedral)

High temperature FeF$_3$
(cubic)

Fig. 1.34. The displacive phase transition observed for ferric fluoride FeF$_3$ when heated above 395°C. It implies just slight rotations of the octahedra.

These notions will be later studied in more detail, once the bases of crystal chemistry are established. Anyway, these transitions have an energetic origin. During the transition, from a minimum of free energy to another, more stability exists in the thermodynamic conditions during the transition.

1.5.5 *The chemical importance of the interstice concept*

A short story about this matter. I remember that one of my students (by the way, brilliant!) came to me at the end of my lecture with the face of somebody who is simultaneously proud to know how to disquiet his professor. Speaking about the lecture, he said that its content was just an exercise of style for testing or reactiviting the knowledge of geometry in the students... I smiled, of course! The young naivety, even pompous, is always pleasant!... I then asked him if he was satisfied with the lithium battery which alimented his cell phone... In front of his disconcerted gaze (probably the sign of his concerns about my mental state!), I was obliged to say him that, if it worked, it was because the interstices in the electrodes of his cell phone were able to host the transferred lithium... A supplementary proof that vacuum must be filled (physically and intellectually as well!). It is known since Aristotle who claimed: *Nature abhors a vacuum*; so, follow him!

In other words, the observation of the interstices in a solid is always a fruitful exercise for the chemist, at two levels: analytical for determining their existence and dimensions, but also creative as soon as their characteristics show the possibility to fill them according to an adapted chemical process. Look at the following example...

For a given type of stacking, the larger the size of the group building this stacking, the larger the empty space between these units. Take the example of the fullerene C_{60}. It corresponds to a *fcc* packing and the tetrahedral and the octahedral interstices are large. For their filling, this would allow to further introduce sodium particles: eight in the octahedral interstices and three in the tetrahedral ones in order to give the solid $Na_{11}C_{60}$. Even if the C_{60} units are no longer tangent, the *fcc* topology is kept!

It is a fruitful idea because, not only was this solid isolated (Fig. 1.35), but it exhibits excellent superconductive properties. This example shows that an idea, purely inspired by crystal chemistry considerations, can be transformed — if we know how to dream! — into an interesting and useful material! This is also the role of the crystal chemist, far beyond the beauty of shapes inspired after so many artists (Paving 1 next page)!

*At this stage of the book, colleagues with some expertise of the complexity of the real solids would experience a few qualms, due to the somewhat boring approach used for the moment. It can become more elegant if some simple **notions of symmetry** are introduced now (they are not absolutely necessary but however recommended). They are developed at the minimum level in Chapter 2. Moreover, the representations of the present chapter correspond to perspective views. They have however limits, those of clarity. Later, another mode of representation must be used: the **projections** on a plane of three-dimensional structures. The figures may seem simpler but need to be read them correctly. This will be developed in Chapter 3.*

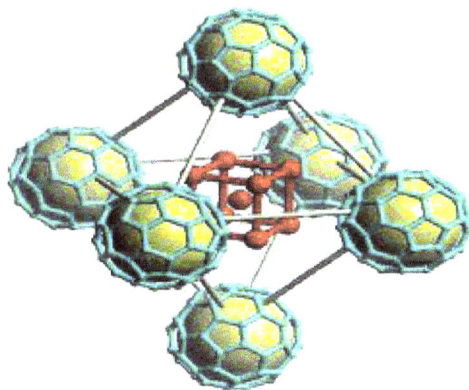

Fig 1.35. Illustration of the insertion of a cube of 8 sodiums in the octahedral interstice of fullerene C_{60}.

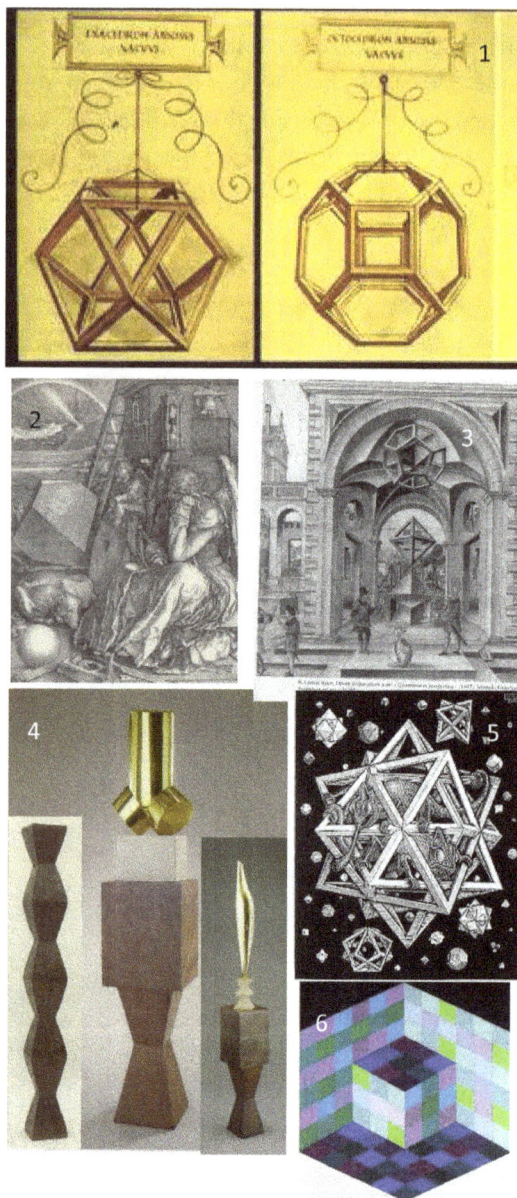

1. Luca Pacioli and Léonard de Vinci:*De divina proportione* (1494)
2. Albert Dürer *Melencolia* (1502)
3. Ludwig Stauer *Geometria* (1567)
4. Constantin Brancusi: *The infinite column* (1937) with some polyhedral pedestals of statues.
5. Mauritius Escher *Stars* (1948)
6. Viktor Vasarely *cube* (1950)

Paving 1.   Whatever the centuries, polyhedra inspired artists with their own sensibility…

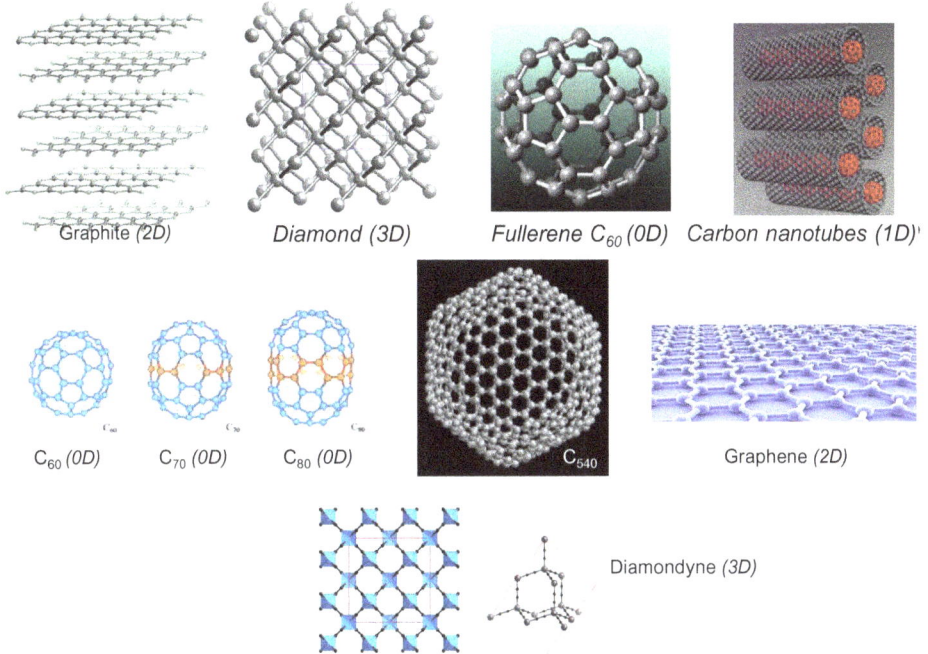

Graphite (2D)   Diamond (3D)   Fullerene C_{60} (0D)   Carbon nanotubes (1D)

C_{60} (0D) C_{70} (0D) C_{80} (0D) C_{540} Graphene (2D)

Diamondyne (3D)

Paving 2. Example of allotropic varieties of an element: the structures of carbon. Note that for fullerene, C60, C70, C80, noted (0D), the corresponding *objects* are zero-dimensional, but their *assembly* by van der Waals bonds is three-dimensional.

Chapter 2

Some Notions of Symmetry

The highest forms of Beauty are order,
symmetry and precision.
Aristotle

The concept of symmetry is absolutely necessary for learning crystallography. Theoretically, that is not the case for a crystal chemist who is *a priori* mainly interested in the arrangement of shapes. As already mentioned, one could ignore this notion in this introduction. It is nevertheless recommended to have in mind some basic notions regarding symmetry for a better understanding of what follows.

Intuitively, it may be thought that the origin of symmetry arrived very quickly after the apparition of *Homo Sapiens* on earth. He looked at the world around him. The reflection of a landscape on the calm surface of the water of a lake gave him the notion of horizontal symmetry. Standing in an upright position and on observing his body, he became conscious of vertical symmetry (Fig. 2.1).

Centuries later — may be! — after his observation of minerals — Nature helped him to discover another type of symmetry: the axial symmetry, which is derived from the observation of regular polygons.

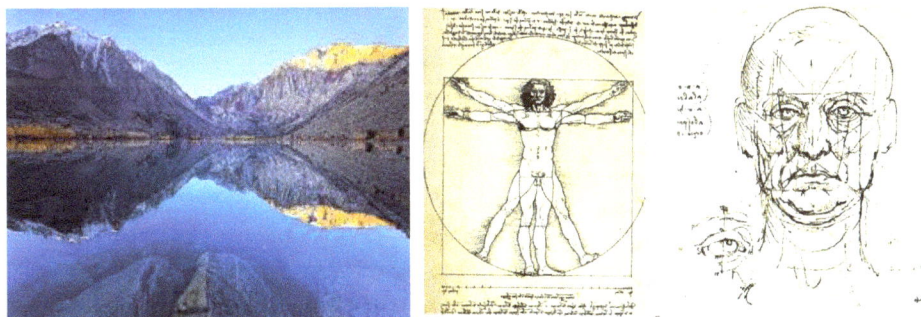

Fig. 2.1. (Left) A lake and horizontal symmetry; (right) early studies on *the man of Vitruve* by Leonardo da Vinci on the proportions of the body and its vertical symmetry.

2.1 The Direct Axial Symmetries

They are associated with the notion of rotation. Indeed, when an object is turned by a certain angle $2\pi/n$, it is found identical to itself. Its initial position is reached after n rotations. This is called symmetry or axis of order n, noted A_n. Figure 2.2 shows the example of the equilateral triangle (symmetry of order 3 or, more simply, symmetry 3).

Intellectually, any type of axis of order n is possible (Fig. 2.3). Only those corresponding to $n = 2, 3, 4, 5$ and 6 are encountered in the three-dimensional space. The value $n = 1$ corresponds to the identity operation. The value $n = 5$ represents a special case, detailed later.

In their current definition, all these operations represent *local* symmetries, which are found in liquid and gaseous molecular species. For their macroscopic

Fig. 2.2. Every $2\pi/3$ rotation of the equilateral triangle gives a figure identical to the initial one. If the red point was absent, the two figures would not be distinguishable.

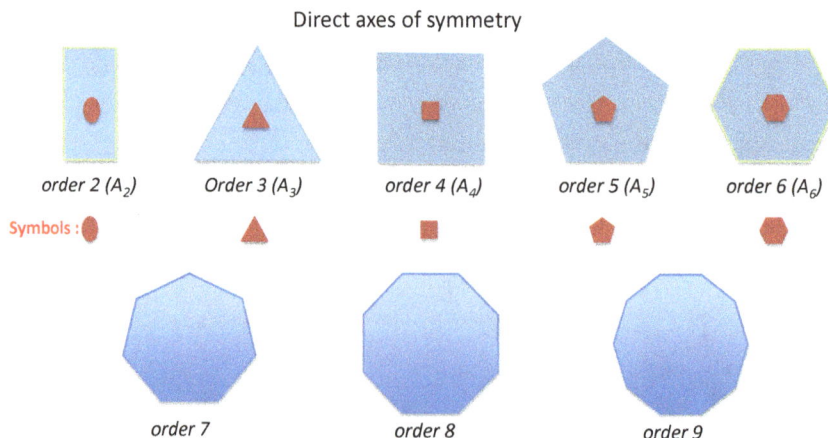

Direct axes of symmetry

order 2 (A_2) Order 3 (A_3) order 4 (A_4) order 5 (A_5) order 6 (A_6)

Symbols :

order 7 order 8 order 9

Fig. 2.3. Representation of regular polygons with their corresponding axes. In the first row, the symbols which appear in red correspond to international codes. Incidently, note that, as soon as regularity disappears, the axis no longer exists.

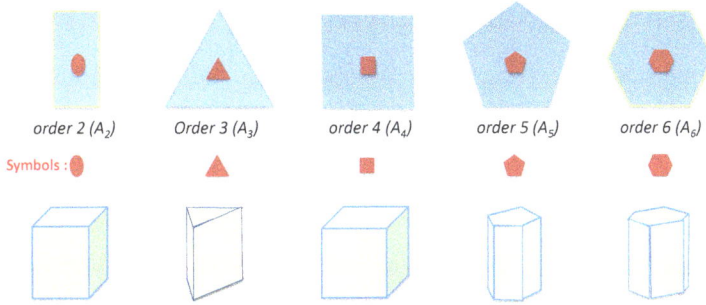

Fig. 2.4. Representation of regular polyhedra with their symmetry axes perpendicular to their polygonal base.

expansion in the solid, it is necessary to add periodicity. At this scale, these initial local symmetries do not appear, and are replaced by others, more general, which concern the infinite solid. Take the example of benzene: in its liquid or gaseous state, a A_6 axis exists for the individual molecule; in solid benzene, even if each molecule keeps its local A_6 symmetry, the three-dimensional mutual arrangement in the solid corresponds to a symmetry of order 2. This has an experimental consequence: the local symmetry is evidenced by spectroscopies, whereas the global symmetry is obtained from diffraction studies (X-rays, neutrons, electrons).

What was true for the plane is also true in the 3D space, and therefore for an isolated cell (Fig. 2.4).

It was necessary to wait for the end of the 19th and the 20th centuries for proving that. At the nanoscale, only 2, 3, 4 and 6 axes can generate the three-dimensional space. In a 2D or 3D space, the five-fold symmetry cannot completely fill the space (Fig. 2.5). Spaces with dimensions >3 are necessary for describing solids with five-fold symmetry, such as quasi-crystals [Daniel Schechtman, Nobel Prize 2011]).

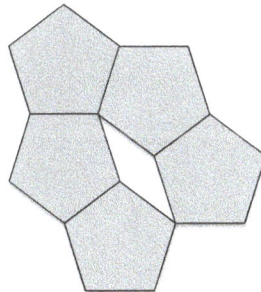

Fig. 2.5. Illustration of the impossibility to realize the complete covering of a plane by identical regular pentagons sharing edges.

2.2 The Inverse Axial Symmetries

When starting from direct axes, they are obtained by adding a center of symmetry belonging to the axis. They are denoted $A_{\bar{n}}$.

Figure 2.6 shows the corresponding operations. Some of them (−2 and −6 axes) are reducible to operations or the combination of direct operations. The −2 axis is equivalent to a mirror perpendicular to the direction of the axis. The −6 axis is equivalent to the combination of a 3 axis and a mirror perpendicular to it; −3 and −4 axes correspond to intrinsic operations.

In Chapter 1, were described the seven crystal systems. Like for dense packings, it was shown that the cell can be primitive (symbol **P**), centered (symbol **I**), or face centered (symbol **F**). The other type of lattice, characterized by only two faces centered (symbol **C**), but which respects the criteria of axial symmetries, is only encountered in monoclinic and orthorhombic systems.. The 14 resulting combinations are known as the **Bravais lattices** (Fig. 2.7).

The French Auguste Bravais (1811–1863), was simultaneously a sailor, an astronomer, a physicist, a mineralogist and a geologist. His numerous achievements concern his observations in geology, his works on crystalline lattices in mineralogy, on atmosphere and optical phenomena at the seaside in geophysics. He was elected to the French Academy in 1854.

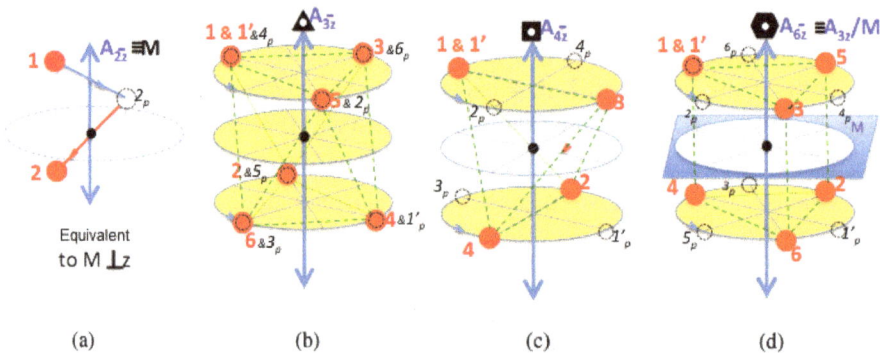

Fig. 2.6. Determination of the equivalent positions generated by the inverse axes (a) −2, (b) −3, (c) −4 and (d) −6. Centers of symmetry are represented as small black spheres at the center of each figure. In each case, these positions appear as red spheres, numbered following their order of apparition during the operation. The dotted circles represent the intermediary position corresponding to the rotation around the direct axis before the center of symmetry operation. If the initial atom is numbered **x**, the intermediary position is noted $(x + 1)_t$ (t for temporary). It is from this that the equivalent sphere (noted **x + 1**) is obtained. The entire operation is finished when the obtained equivalent position corresponds to the starting position **1**. Incidentally, note that the −3 axis generates the vertices of an octahedron (green dotted lines). In the same way, the −4 axis generates the vertices of a tetrahedron, whereas the −6 axis defines a triangular-based prism.

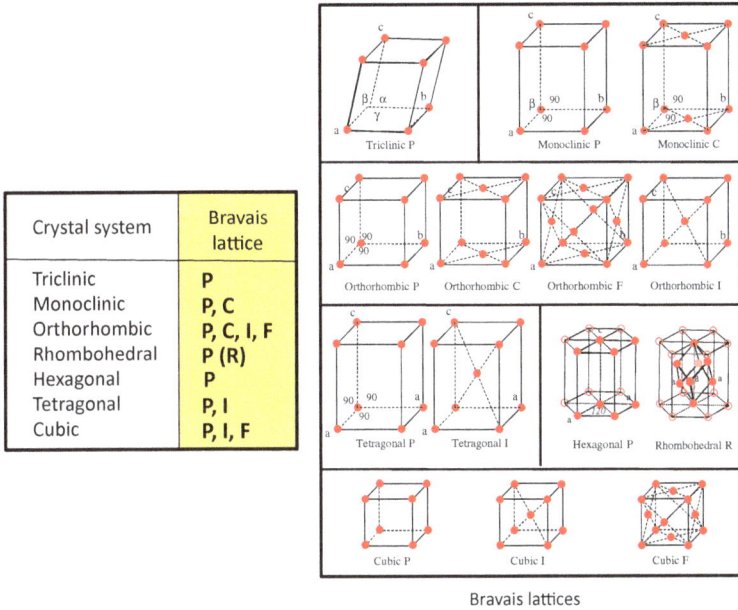

Crystal system	Bravais lattice
Triclinic	P
Monoclinic	P, C
Orthorhombic	P, C, I, F
Rhombohedral	P (R)
Hexagonal	P
Tetragonal	P, I
Cubic	P, I, F

Bravais lattices

Fig. 2.7. The 14 Bravais lattices and their illustration.

When a complex group is submitted to an operation of axial symmetry, all its constitutive elements are also submitted to the same operation, as shown in Fig. 2.8 for some examples.

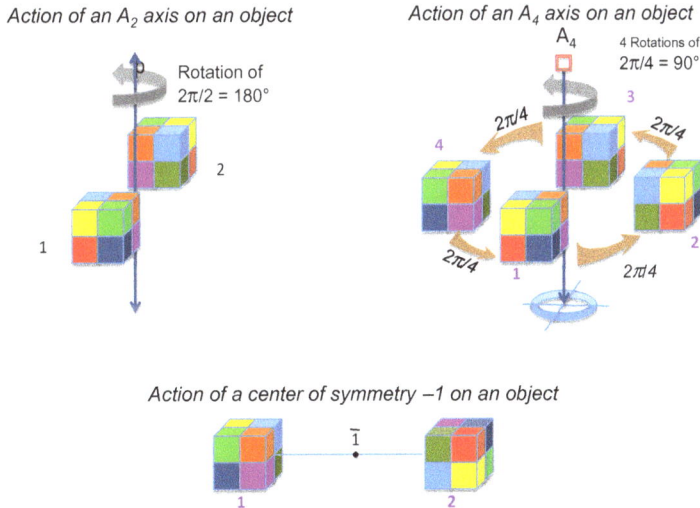

Action of an A_2 axis on an object

Rotation of $2\pi/2 = 180°$

Action of an A_4 axis on an object

A_4 4 Rotations of $2\pi/4 = 90°$

Action of a center of symmetry −1 on an object

$\bar{1}$

Fig. 2.8. Some examples of axial symmetry operations on a complex object (represented here by the assembly of cubes of different colors, and forming the complex object, everytime noted 1 at the beginning of the operations).

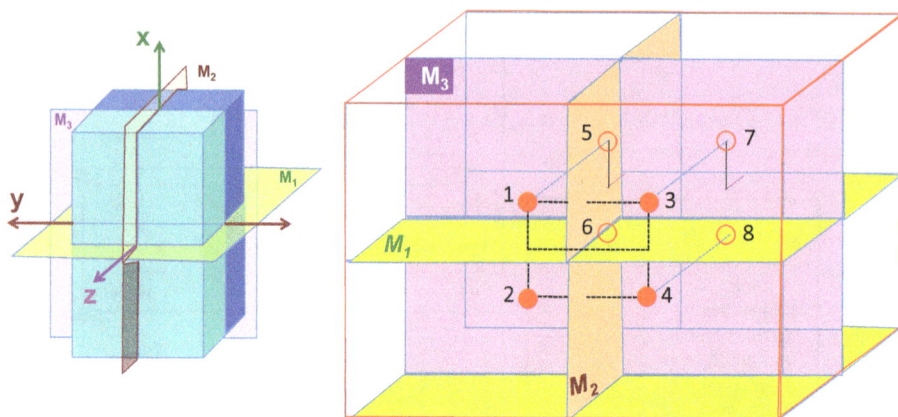

Fig. 2.9. (Left) Visualization of the three perpendicular mirrors existing in a primitive orthorhombic cell and of the elements of symmetry induced by their existence (see text); (right) determination of the equivalent positions (numbered from 2 to 8) deduced from the general position 1 (x, y, z) by the successive operations of the three mirrors.

Within the same cell, a combination of axial symmetries and plane symmetries (or mirrors) is possible, but the number of combinations remains limited and restricted to only thirty two possibilities in a 3D space. They are tabulated and called the **32 point groups**.

Take the example of an orthorhombic cell ($a \neq b \neq c$; $\alpha = \beta = \gamma = 90°$). In it, three perpendicular mirrors, noted M_1, M_2 and M_3, can be evidenced (Fig. 2.9).

Take an atom (noted 1) within the cell, outside any symmetry element. In crystallographic language, its position is called *general*. Let it be submitted successively to the symmetry operation corresponding to a mirror. The operation M_1 generates position 2 from 1. The operation M_2, applied to both 1 **and** 2, gives 1→3 and 2→4. The operation M_3, this time applied to 1, 2, 3 and 4, leads to 1→5, 2→6, 3→7 and 4→8. This means that, in this primitive orthorhombic cell, there are **eight positions equivalent to that of the initial 1**. It is worthy to note that, if 1 was located on a mirror, it would have created only four equivalent positions. In the same way, placed at the intersection of two mirrors, only two positions would have been generated, and only one (itself) when it is located at the intersection of the three! Therefore, a first lesson on this point: For every system of symmetry operators, the number of positions equivalent to that of a starting position depends on the place of the latter in the cell.

Moreover, as all these positions are equivalent, one can pass in every case from one to the other, whatever the nature of the operator. Therefore, what operator ensures the passages?:

- 1→4, 2→3, 5→8 and 6→7? Clearly a two-fold axis parallel to the *x* direction: $\mathbf{A_{2x}}$!
- 1→6, 2→5, 3→8 and 4→7? a two-fold axis parallel to the *y* direction: $\mathbf{A_{2y}}$!
- 1→7, 2→8, 3→5 and 4→6? a two-fold axis parallel to the *z* direction: $\mathbf{A_{2z}}$!

Finally, the possible passages 1→8, 2→7, 3→6 and 4→5 prove the presence of a **center of symmetry** at the center of the cell. In other words, the three mirrors acting as symmetry operators in the orthorhombic cell **induce** other symmetry elements: three two-fold axes and one center of symmetry. They appear as in Fig. 2.9 (left).

Why such inductions? Look this time at the example of Fig. 2.10. It corresponds to either a mirror and a perpendicular two-fold axis (A_{2z}), or two perpendicular A_2 axes (A_{2y} and A_{2z}). Both lead to an identical result: the induced generation of a symmetry center.

This result is valid whatever the nature of the A_{nz} axis ($1 < n \le 6$).

General rule: For a given crystal system, whatever the choice of the starting elements of symmetry, the symmetry elements induced by the choice must be investigated.

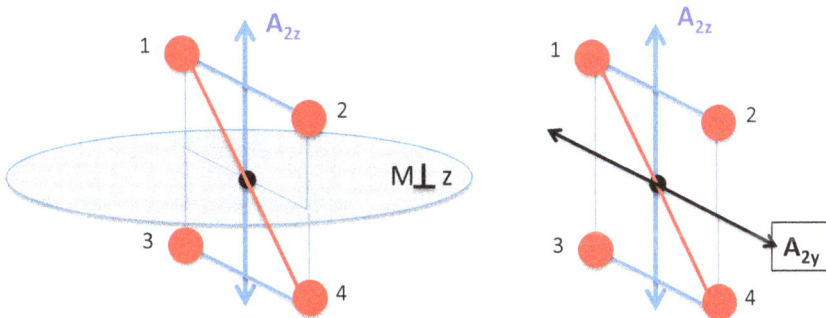

Fig. 2.10.   (Left) Equivalent positions of a general position 1 by the joined action of a two-fold axis (A_{2z}) and of a perpendicular mirror *M*; (right) or of two perpendicular two-fold axes (A_{2y} and A_{2z}). In both cases, this double action induces the existence of a center of symmetry (black small circles).

2.3 The Symmetries at the Atomic Scale

Symmetries coexist with those existing at the macroscopic scale. Intrinsic to the atomic order, they concern ***helicoidal (or screw) axes*** and ***glide planes***.

An *helicoidal axis* (noted An_p or n_p axis) corresponds to two successive operations applied to a position in the cell: (i) first, a $2\pi/n$ counter clockwise rotation around the axis, followed by (ii) a translation of a fraction p/n of the parameter parallel to the considered axis. Figure 2.11 details the succession of these operations for a 2_1 axis.

Figure 2.12 consolidates this notion. It details the succession of the operations associated with a 4_1 axis, leading to the four equivalent positions of an initial one. Note that the symbol of the 4_1 axis indicates the counterclockwise sense of rotation.

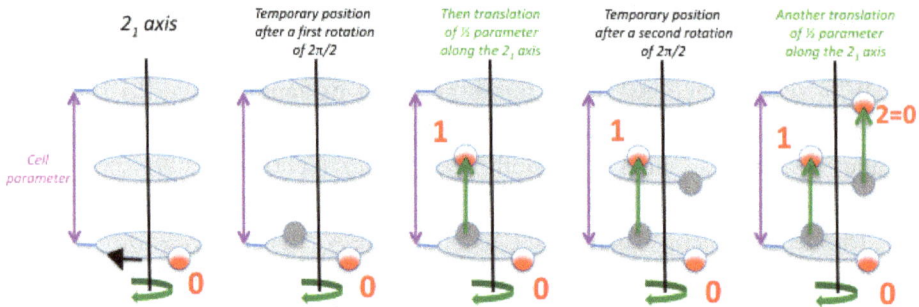

Fig. 2.11.   Details of the succession of operations generating the equivalent position of an initial one through a 2_1 axis operator.

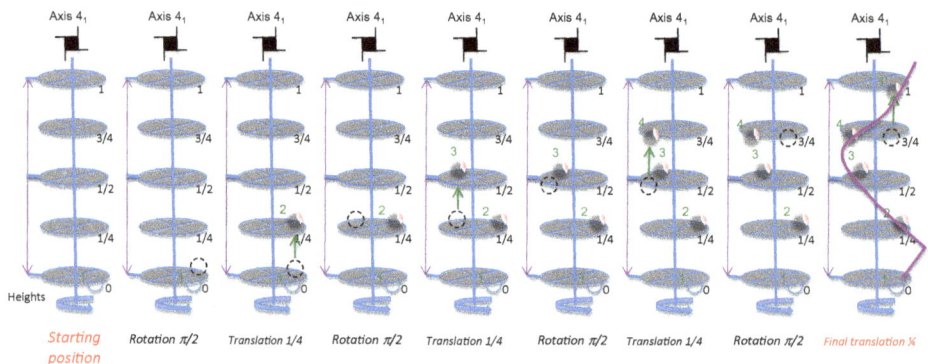

Fig. 2.12.   Details of the succession of operations generating the equivalent positions of an initial one through a 4_1 axis operator. Their positions follow an helix (in purple in the figure on the right). The purple vertical arrows represent the cell parameter parallel to the 4_1 axis. The green arrows illustrate the successive translations of ¼ of the parameter.

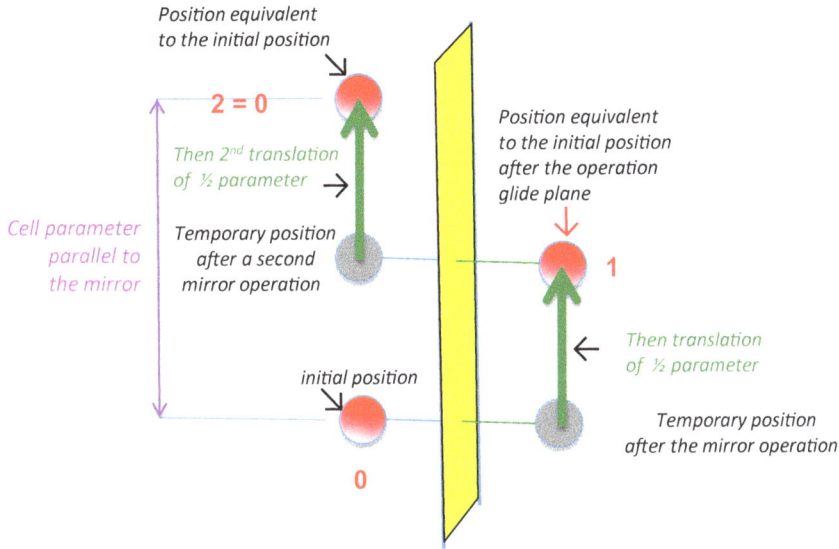

Fig. 2.13.   Details of the successive operations generating the equivalent position of an initial one through the action of a glide plane.

In the *glide plane* operation (Fig. 2.13), after the classical mirror operation, a translation in one or two directions parallel to the mirror follows. The modulus of each translation vector is equal to one-half of the cell, the parameter being in the considered direction.

In terms of notation, instead of being noted *m* as in "classical" planes of symmetry, these glide planes are written in bold italics according to the direction of the translation vector: *a* if the translation of ½ parameter is parallel to axis *a* of the referential, *b* if the translation is parallel to axis *b*, *c* if the translation is parallel to axis *c*. If, after the mirror operation, the translation successively occurs along two axes parallel to this mirror, the glide plane is noted *n*. Note that, sometimes, *d* glide planes also exist. They correspond to two successive translations: the first, parallel to the plane, with modulus of ¼ of the corresponding parameter, the second, also with modulus ¼ of the corresponding parameter, this time in a direction perpendicular to the same plane.

It was previously shown that "pure" axes and mirrors could be combined. It is the same for screw axes and glide planes. The number of possibilities of combination is by far larger. Instead of the 32 classes of symmetry already mentioned, it is now **230**. They are called the ***230 space groups***.

Figure 2.14 illustrates the combined action of a 2_1 axis parallel to the crystallographic axis *b* and of a glide plane *c* (which means a translation of $\frac{1}{2} \cdot c$ after the mirror operation), perpendicular to the 2_1 axis along *b*.

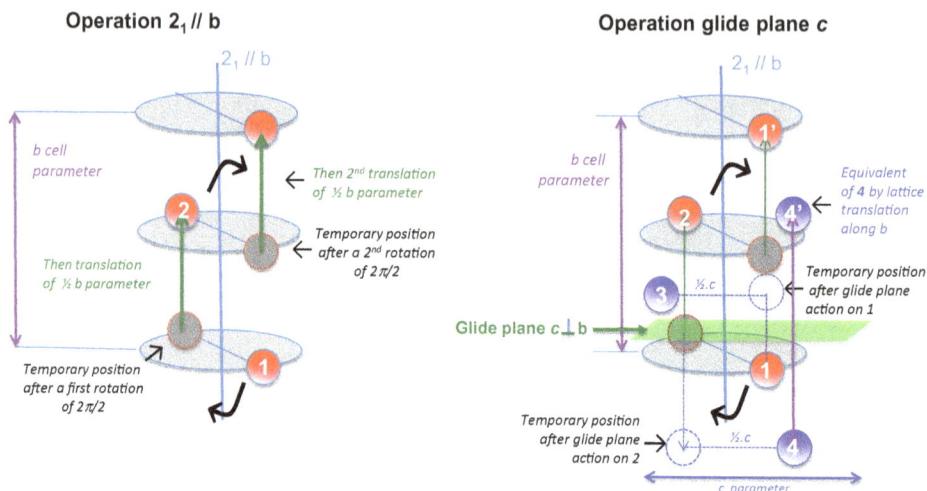

Fig. 2.14.   Details of the succession of operations generating the equivalent positions of an initial one through successive actions of (left) a 2_1 axis and then (right) of a glide plane **c**.

Beginning from the 2_1 operator on position 1, 1 → 2. The action of the glide plane applied to positions 1 and 2 then gives 1 → 3 and 2 → 4 after temporary positions due to the "pure" mirror operation (dotted lines) on which the translation of $\frac{1}{2} \cdot c$ is applied for each of them, leading to the 3 and 4 positions. Note that the 4 equivalent position is primitively outside the cell but, due to the periodicity, the same position (4′) also exists within the cell. It is obtained by a translation, the modulus of which being the same as the modulus of the cell parameter *b* in this direction.

Figure 2.15 provides the international symbols of screw axes and glide planes, according to their direction reported from the plane of projection.

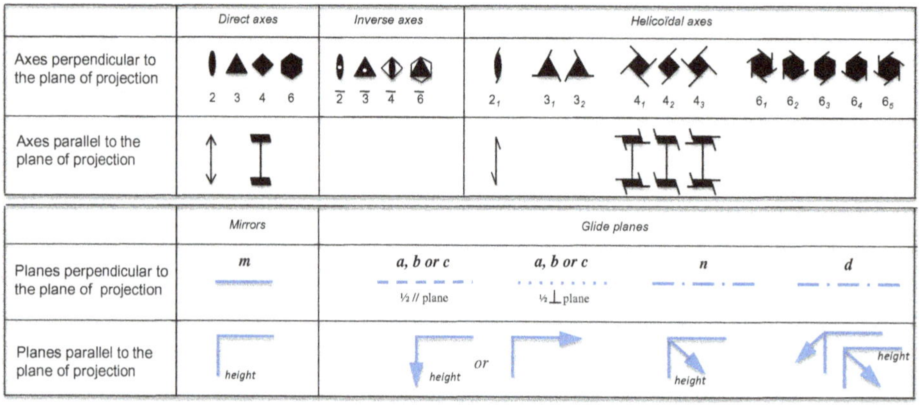

Fig. 2.15.   International symbols for the various symmetry elements.

Incidently, whereas Fig. 2.14 corresponds to a very simple case, it becomes obvious that it will be more and more difficult to represent a perspective view of the equivalent positions in the space. The drawing rapidly becomes confusing, even unreadable. Simplification of the representation then requires to work on projections of the structures, even if their interpretation will become more difficult for some of the readers. The present chapter will provide an initiation to them. It will be more deeply developed in Chapter 3 and applied to more complex structures.

Take for example a primitive monoclinic cell (lattice *P*) [$a \neq b \neq c$; $\alpha = \gamma = \pi/2$; $\beta \neq \pi/2$], with the 2_1 axis at 0; *y*; ¼ and the perpendicular glide plane *c* at the reduced level ¼ (reduced level: the fraction of the *b* parameter at which the glide plane is situated). This is the convention decided by the *International Union of Crystallography*, and the corresponding space group is noted **P 2₁/c**. The initial position **1**, whose reduced coordinates are *x*, *y* and *z*, is first submitted to the 2_1 operation (axis parallel to *b*). Its height along *y* will be noted + on the projection [Fig. 2.16(a)].

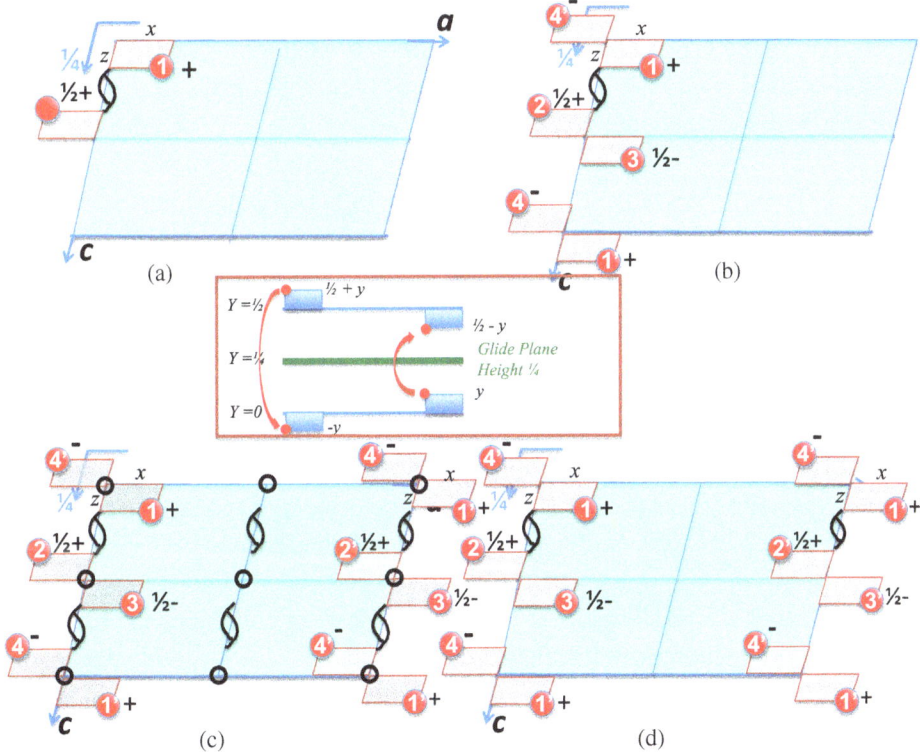

Fig. 2.16. Projection of the successive steps leading to all the symmetry elements present in this space group. The symmetry centers appear as black empty circles and the equivalent positions as filled red circles at the center of which the number characterizes the position. The central inset explains the height changes resulting from the glide plane operation.

Constrained by the position of the 2_1 axis, the transformation of coordinates from **1** to **2** gives for **2** the values $-x$, $\frac{1}{2} + y$, $\frac{1}{2} - z$ (x becomes $-x$, y becomes $\frac{1}{2} + y$ and z becomes $\frac{1}{2} - z$).

After the glide plane operation, **1** → **3** and **2** → **4** (*and **4′** by lattice transla-tion*). Due to the position of **c** plane at the height ¼, the reduced coordinates of atom **3** are such that x remains x, y becomes $\frac{1}{2} - y$ and z becomes $\frac{1}{2} + z$. For **4**, x becomes $-x$, y becomes $-y$ and z becomes $-z$.

The combination of the operations 2_1 and **c** therefore generates four equivalent positions with coordinates $x\ y\ z$; $-x$, $\frac{1}{2} + y$, $\frac{1}{2} - z$; x, $\frac{1}{2} - y$, $\frac{1}{2} + z$; $-x\ -y\ -z$ or, in a more condensed manner, $\pm(x\ y\ z;\ x,\ \frac{1}{2} - y,\ \frac{1}{2} + z)$. This characterizes the **P 2_1/c** space group in which **P** relates to the Bravais lattice and 2_1**/c** indicates that 2_1 is perpendicular to the glide plane **c**.

The induced symmetry elements are easily accessible from the projection. If the passages **1** → **2** through 2_1 and **1** → **3** and **2** → **4** through **c**, were explained by these operators, how can we explain the passages **1** → **4** and **2** → **3**? It is clear that it is through a **center of symmetry "O"**. With this, the space group becomes complete for generating all the equivalent **general** positions. But these positions do not exist alone. If 1 is not in a general position (which means outside any opera-tor), it occupies what is called a particular position which leads to fewer equiva-lents. In the **P 2_1/c**, they are:

$$0\ 0\ 0, \quad \tfrac{1}{2}\ 0\ 0, \quad 0\ 0\ \tfrac{1}{2}, \quad \tfrac{1}{2}\ 0\ \tfrac{1}{2},$$
$$0\ \tfrac{1}{2}\ \tfrac{1}{2} \quad \tfrac{1}{2}\ \tfrac{1}{2}\ \tfrac{1}{2} \quad 0\ \tfrac{1}{2}\ 0 \quad \tfrac{1}{2}\ \tfrac{1}{2}\ 0.$$

In other words, using a logical and simple way, it was possible to completely iden-tify a space group. Fortunately, it is not necessary to determine every time a given space group. Now, the characteristics of the 230 groups are tabulated and described in an important book, ***The International Tables for Crystallography***, established by the International Union of Crystallography. In the book, a double page is dedi-cated to each group: on the left page, appear the projections along the three axes; the right page gives the coordinates of all the possible general and specific posi-tions (Fig. 2.16).

The letters h, k and l in italics appearing in the last column of the right page are called **Miller indices**. When they are in parentheses: (hkl), this characterizes a family of parallel planes and only one; when in brackets: $[hkl]$, they indicate a direction **perpendicular** to the (hkl) plane.

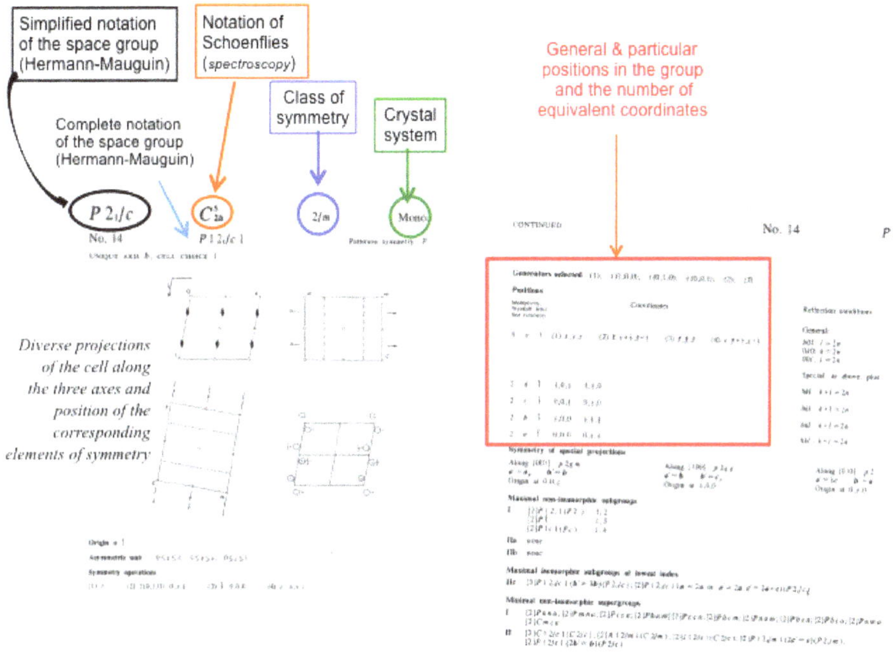

Fig. 2.17. Information contained in the double page describing a space group (here *P2₁/c*). In the red frame, the first column gives in decreasing order the number of equivalent positions, general and specific. The letters of the second column, associated with the number of equivalents, follow the alphabetic order from down to top. For crystallographers, the association (number + letter) characterizes one type of crystallographic site. For instance, the general position is named (4e) site. This notation, used worldwide is due to R.W.G. Wyckoff (an American crystallographer (1897–1994)). The third column gives the point of symmetry of the corresponding Wyckoff site (here 1 means that the site is outside any symmetry element; −1 indicates a center of symmetry). The other columns detail the reduced atomic coordinates of each equivalent positions of the given site outside.

2.4 The Miller Indices

Consider a general trihedron OXYZ, its three unit vectors **a**, **b**, **c** and a plane, closer to the origin but not containing the origin. It cuts the three axes at distances *m***a**, *n***b** and *p***c** from the origin, *m*, *n* and *p* being in the interval [−1,1]). *h*, *k* and *l* respectively correspond to 1/*m*, 1/*n* and 1/*p*. They define the **Miller indices** of a whole family of planes (*hkl*) parallel to the considered initial plane, defined above. The distance between each of them is the same as the distance between the origin and the considered initial plane. ***h, k and l are integers***. The line perpendicular to this family of planes correspond to the direction noted [*hkl*]. Note the importance of the sentence "closer to the origin". For instance, if $m = n = p = \frac{1}{2}$, the plane will be (222) and if $m = n = p = 1$, the plane will be (111). A family of parallel planes are characterized by a single series of indices *h,k,l*, which are numbers prime to one another in a primitive cell. Figure 2.18 gives some examples of frequently encountered planes and directions with their Miller indices.

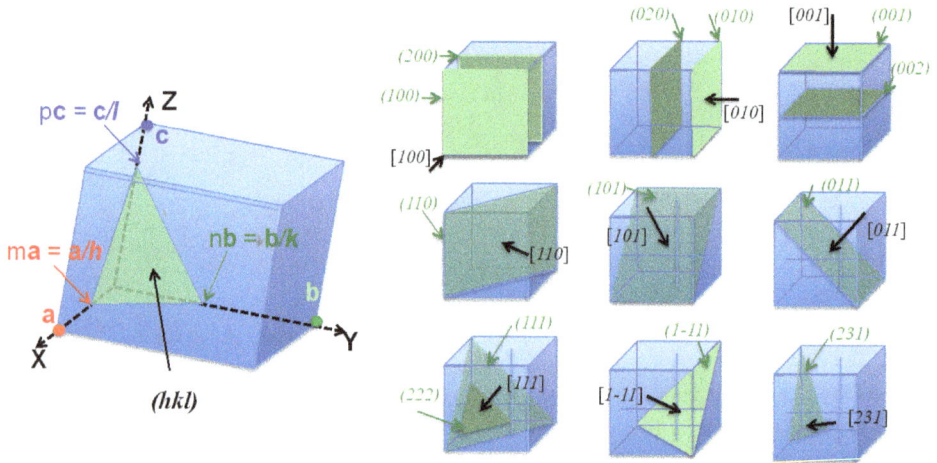

Fig. 2.18. (Left) Definition of the Miller indices; (right) some examples of planes (green) and directions (black arrows) with their Miller indices.

Chapter 3

The Projections, Their Reading, Their Uses...

Projection? Flat drawing or jump in the future?
Linus Pauling (*How to live longer and feel better*)

In the previous chapter, it was mentioned that, as soon as the complexity of a structure increases, perspective views become more and more unreadable. It therefore becomes necessary to work in two dimensions (2D) instead of three. With a minimum of care, the drawing of the projection is easy but, on the contrary, its interpretation becomes more difficult. It is not always evident, but there is a chapter for learning how to "read" a structure.

3.1 The Steps of the Examination

3.1.1 *The drawing*

As seen before, after his/her determination of a structure, the crystallographer publishes a table containing:

— the formulae of the compound and, eventually its name (if the solid is a mineral),
— the crystalline system, the cell parameters and the space group,
— the reduced coordinates of all the independent crystallographic sites, with the name of the element occupying this site.

If the chemist is also a crystal chemist, he will know how to decipher this table in order to transform the numerical data of the table into a drawing, as clearly as possible, for visualizing the atomic arrangements of atoms in the 3D space. This is the first step in the work of the crystal chemist, impossible to circumvent.

The first example developed in this chapter concerns rutile, a natural mineral, which is one of the varieties of titanium dioxide TiO_2.

The table of crystal data is the following:

TiO$_2$ rutile form	space group: P 4$_2$/mnm (n° 136)	
Tetragonal	$a = 4.5937$ Å	$c = 2.9581$ Å
Ti 2a	0 0 0	
O 4f	$x x$ 0 with $x = 0.3053$	

Rather short! This only means that titanium (position 2a) occupies two equivalent positions, and oxygen (position 4a) has four. Going to the double page of space group n° 136 of the *International Tables* gives the coordinates of all these positions:

2a: 0 0 0 [Ti$_1$] and ½ ½ ½ [Ti$_2$]

4f: ($x x$ 0 [O$_1$], $-x -x$ 0 [O$_2$], ½ + x, ½ – x, ½ [O$_3$] and ½ – x, ½ + x, ½ [O$_4$].

Oxygen atoms being in a general position, with $x = 0.3053$, a perspective view is not recommended. A projection must be preferred. The best choice for the plane of projection is the (a, b) plane, which is perpendicular to the screw axis 4$_2$, which is the projection axis.

Working in projection implies to write beside each atom its height along the z axis. This leads to Fig. 3.1(a) which represents the atoms within the tetragonal cell. Further lattice translations applied to the six atoms of the cell lead to Fig. 3.1(b).

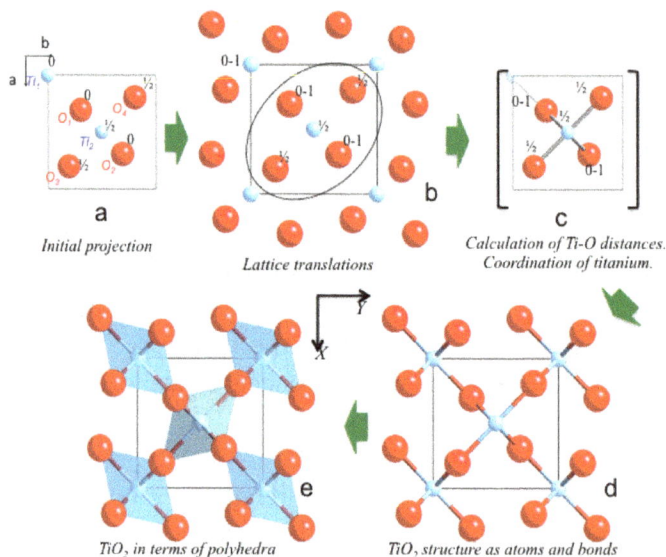

Initial projection

Lattice translations

Calculation of Ti-O distances.
Coordination of titanium.

TiO$_2$ in terms of polyhedra

TiO$_2$ structure as atoms and bonds

Fig. 3.1.   The various sequences of the visualization of the (a,b) projection of the TiO$_2$ structure: (a) Place the atoms with their different heights within the cell; (b) Extension to other cells by lattice translations applied to each of the six atoms of the initial cell; (c) Ti-O distance calculations for evaluation of the nature of titanium environment; (d) the (a,b) (or (001)) projection of the TiO$_2$ structure including the Ti-O bonds; (e) rutile represented as an association of octahedra.

It becomes qualitatively clear that the Ti^{4+} cation at the height ½ is surrounded by six oxygen atoms: The calculation of the various Ti-O distances is therefore necessary: two oxygen atoms at the same height ½, two oxygen atoms at $z = 0$, and two others at $z = 1$. The line joining these four latter oxygen atoms ($z = 0$ and 1) on the projection form a square around Ti^{4+} for a perspective view, and the line joining the two oxygens at $z = $ ½ seems to be perpendicular to the square. Qualitatively, it seems that Ti^{4+} is in a six-fold coordination. This has to be proved quantitatively.

3.1.2 *Calculations of distances and coordinations*

Even if helpful, computer programs of drawing the structures exist now, which provide by simple clicks the values of the distances between two atoms and of the angles between three atoms, it seems necessary to recall to the reader the classic basic calculations for obtaining these values. It is always essential to understand what one does... The danger of these helpful current programs, which are however just "black boxes", is to forget the fundamentals...

A mathematical reminder: the distance between two points in a general trihedron.

In an classical direct trihedron (OXYZ), defined on the three axes by the unit vectors *a*, *b*, *c* (which are the parameters of the cell) (Fig. 3.2), points **A** and **B** have for coordinates $(x_A a, y_A b, z_A c)$ and $(x_B a, y_B b, z_B c)$, respectively. The numbers x_A, y_A, z_A and x_B, y_B, z_B represent the **reduced coordinates** of each point (i.e. the fraction of the parameter where the projection of the point arrives). The *vector* **AB** is then defined by the expression $[\Delta x \cdot a + \Delta y \cdot b + \Delta z \cdot c]$ in which the symbol Δ represents the difference between the coordinates **A** and **B** along the considered axis (for instance, $[\Delta x \cdot a = (x_B - x_A) \cdot a]$).

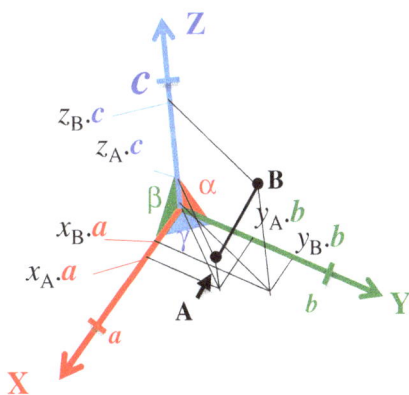

Fig. 3.2. International conventions for the representation of a classical direct trihedron (OXYZ). The unit vectors along the three axes are noted **a, b** and **c.**

The scalar product of the vector by itself represents the square of the modulus of the distance d between $\mathbf{A}$ and $\mathbf{B}$.

$$d^2 = [\Delta x \cdot \mathbf{a} + \Delta y \cdot \mathbf{b} + \Delta z \cdot \mathbf{c}] \cdot [\Delta x \cdot \mathbf{a} + \Delta y \cdot \mathbf{b} + \Delta z \cdot \mathbf{c}]$$

which, by developing this product, gives:

$$d^2 = [\Delta x^2 \cdot \mathbf{a}^2 + \Delta y^2 \cdot \mathbf{b}^2 + \Delta z^2 \cdot \mathbf{c}^2 + \Delta x \cdot \Delta y \cdot ab \cdot \cos \gamma$$
$$+ \Delta x \cdot \Delta z \cdot ac \cdot \cos \beta + \Delta y \cdot \Delta z \cdot bc \cdot \cos \alpha].$$

This general expression can be declined for each of the already described seven crystalline systems. It obviously remains unchanged for the triclinic system and becomes simpler for the other systems.

Monoclinic system $(a \neq b \neq c; \alpha = \gamma = \pi/2; \beta \neq \pi/2)$

$$d^2 = [\Delta x^2 \cdot \mathbf{a}^2 + \Delta y^2 \cdot \mathbf{b}^2 + \Delta z^2 \cdot \mathbf{c}^2 + \Delta x \cdot \Delta z \cdot ac \cdot \cos \beta]$$

Orthorhombic system $(a \neq b \neq c; \alpha = \beta = \gamma = \pi/2)$

$$d^2 = [\Delta x^2 \cdot \mathbf{a}^2 + \Delta y^2 \cdot \mathbf{b}^2 + \Delta z^2 \cdot \mathbf{c}^2]$$

Rhombohedral system $(a = b = c; \alpha = \gamma = \pi \neq \pi/2)$

$$d^2 = [(\Delta x^2 + \Delta y^2 + \Delta z^2 + (\Delta x \cdot \Delta y + \Delta x \cdot \Delta y + \Delta x \cdot \Delta z + \Delta y \cdot \Delta z) \cdot \cos \alpha] \cdot \mathbf{a}^2$$

Hexagonal system $(a = b \neq c; \alpha = \beta = \pi/2, \gamma = 2\pi/3 \ (\cos \gamma = -1/2)$

$$d^2 = [(\Delta x^2 + \Delta y^2 - \tfrac{1}{2} \cdot \Delta x \cdot \Delta y) \cdot \mathbf{a}^2 + \Delta z^2 \cdot \mathbf{c}^2]$$

Tetragonal system $(a = b \neq c; \alpha = \beta = \gamma = \pi/2)$

$$d^2 = [(\Delta x^2 + \Delta y^2) \cdot \mathbf{a}^2 + \Delta z^2 \cdot \mathbf{c}^2]$$

Cubic system $(a = b = c; \alpha = \beta = \gamma = \pi/2)$

$$d^2 = (\Delta x^2 + \Delta y^2 + \Delta z^2) \cdot \mathbf{a}^2.$$

Coming back to the structure of rutile, the use of the formula of the tetragonal system and the values of the coordinates of the different atoms of the cell shows that there are two types of distances d_1 and d_2 around Ti^{4+} when the x value given for oxygen is used: $d_1 = 1.983$ Å and $d_2 = 1.946$ Å [Figs. 3.3(a)–3.3(c)].

The small difference between the two values leads to a general remark. If the values of the distances are strictly applied, Ti^{4+} would have four neighbors, forming a rectangle around it. However, as the two values d_1 and d_2 are extremely close (2% of difference), one can consider with a very good approximation that they are equal. In this case, the coordination number becomes six and the rectangle is transformed into an octahedron. Indeed, the octahedron is not regular: the calculation

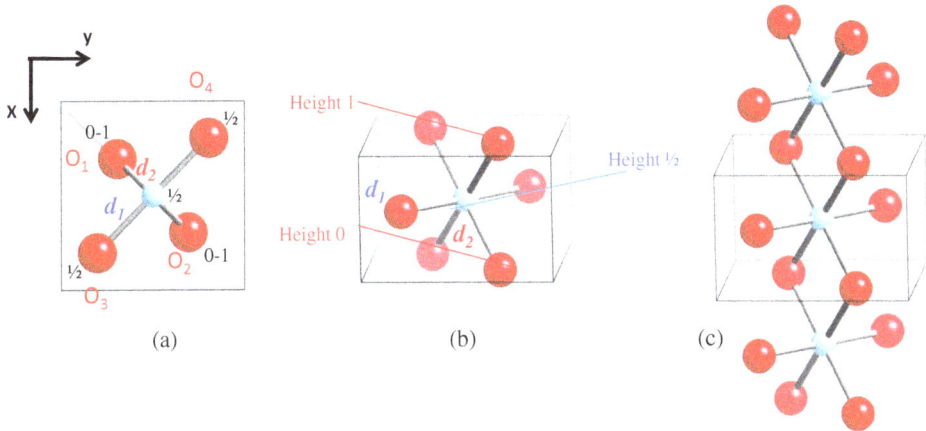

Fig. 3.3. (a) (001) projection of the Ti^{4+} environment; (b) the same with perspective view for a better understanding of the heights; (c) extension along z, showing in this direction, the edge-sharing connection of titanium octahedra.

of the various O–O distances provides three different values: 2.530 Å, 2.779 Å and 2.958 Å (the latter representing the c parameter of the tetragonal cell). Despite the large discrepancy (ca. 30%) between the extreme distances, the coordination is kept as octahedral. This illustrates that crystal chemistry can present some discards from the ideal state when real solids are considered.

The difference between the values of the O–Ti–O angles around the cation is another expression on this non-regularity. The application of the formula related to the ordinary triangle ($a^2 = b^2 + c^2 - 2bc \cdot \cos\alpha$ and circular permutations) leads to the values of angles between bonds. Keeping the notations for the atoms in the rutile structure, the values for the corresponding angles are: O_1–Ti_2–O_1 = 98°93; O_1–Ti_2–O_2 = 81°07; O_1–Ti_2–O_3 = 90°. In the same way, the angle Ti_1–O_1–Ti_2 = 130° — called super-exchange angle — is very important for the prediction of the type of magnetic coupling which will exist between two $3d$ or $4f$ cations between which an anion is inserted.

Incidently, the values of the distances reflect the nature of the chemical bond between titanium and oxygen. From consulting the Shannon's table of atomic radii, it appears that the calculated Ti–O distances are close to the sum of the ionic radii of Ti^{4+} in six-fold coordination and of oxygen in three-fold one. Therefore, the Ti–O bond in rutile is essentially ionic.

From what was shown above, it becomes clear that the concept of coordination, deduced from a crystallographic study, is not strict but only qualitative with "reasonable" error bars. However, what are the limits of "reasonable"? If one excepts the case of Jahn–Teller ions like Cu^{2+}, Cr^{2+} or Mn^{3+}, in which the difference of distances is rather large, due to orbital specificities, it is commonly admitted that distances can be considered as equal if the distances differ by less than 10%. For avoiding this ambiguity, another approach (the bond valence method) will be developed later in this book and illustrated with several examples.

The spectroscopic behavior of ions also provides information on cationic coordination. In courses on general chemistry, the crystal field theory is usually developed. It foresees the changes in degeneracy of the levels of *3d* orbitals (on at least two levels) as a function of the coordination. If an ambiguity remains after crystal chemistry calculations (but it is rare!), a simple spectroscopic measurement gives the answer.

3.1.3 *Connections of polyhedra*

It was shown that, in rutile, Ti^{4+} adopts an octahedral coordination. If lattice translations are applied to the central octahedron inside the cell, two other octahedra appear above and below the starting octahedron [Fig. 3.3(c)], and share edges with it. This means that the Ti^{4+} octahedron at height ½ of Fig. 3.3 corresponds in reality to infinite columns of edge-sharing octahedra which develop along the *c* axis of the rutile cell.

What happens for the titaniums located at 0 and 1 levels? Looking at Fig. 3.1(e), it is obvious that, for them, the same situation occurs, with however two notable differences: the height of Ti^{4+} and the orientation of the octahedra, perpendicular to the first column. It is more visible when two different colors are used [Figs. 3.4(a) and 3.4(b)]. Figure 3.4(b) clearly shows that the connection between the two types of columns occurs through vertices. This facilitates the illustration of the perspective view of rutile [Fig. 3.4(c)].

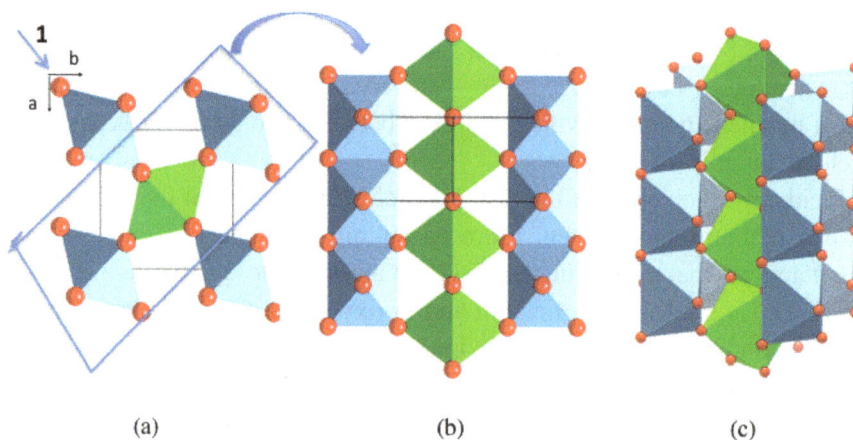

(a)   (b)   (c)

Fig. 3.4. (a) Projection of the rutile structure using two different colors for the two types of columns; (b) view of the structure along [110], viewed along the direction of the arrow 1; it shows better the vertex-sharings between the two columns; (c) perspective view of the rutile structure.

3.1.4 *Dense packings and complex structures*

Up to now, the attention of the reader was focused on the small cations, their environment and the connections of their coordination polyhedra. This allowed a good understanding of the rutile structure. Good but not complete! These small cations are surrounded by large anions (here oxide anions O^{2-}), and it is these anions which are mainly responsible for the entire organization of the structure. It is therefore necessary to also look at this organization, and projections along other directions of the structures.

For the rutile structure, another type of observation is needed with a different eye. Even if Fig. 3.5(a) is the same as Fig. 3.4(a) ((001) projection), attention is paid this time to oxygen. At heights $y = \frac{1}{4}$ and $y = \frac{3}{4}$, they form ondulated (010) planes parallel to XOZ and deserve a special attention. Indeed, the projection along Y (direction [010]) [Fig. 3.5(b)] leads to a distorted hexagonal arrangement of the oxygens. It would be ideal if x (the variable parameter of $4a$ Wyckoff positions) was 0.25 instead of 0.3053.

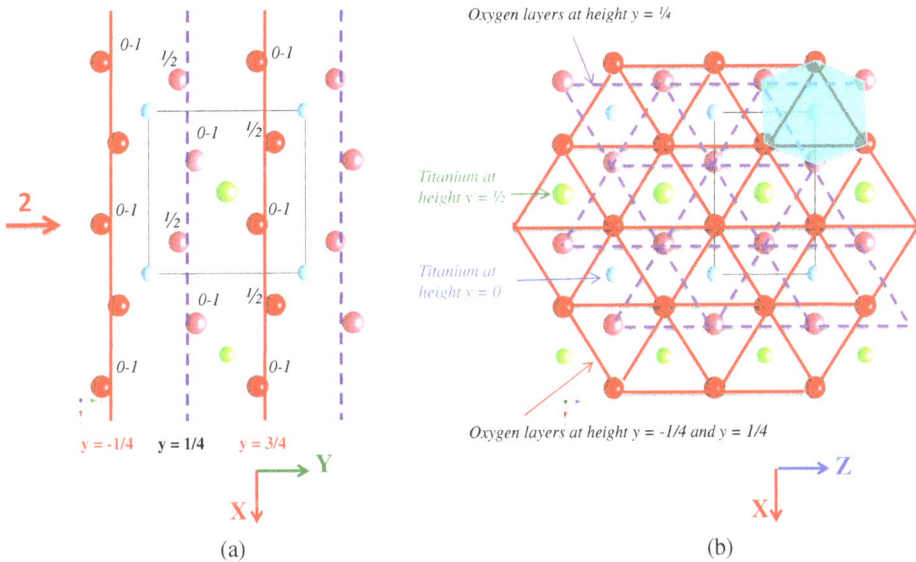

(a)   (b)

Fig. 3.5. Other visions of rutile: (a) (001) projection on the plane XOY, on which atoms are represented by their color and their heights. A periodic succession of two types of corrugated planes appears: type 1 (red continuous lines) and 2 (purple dotted lines); (b) looking along the direction [010], noted as 2 in figure (a) with an arrow, it is clear that the two types of corrugated planes form triangular lattices, in such a way that an atom of type 2 plane projects at the center of a triangle of type 1 atoms. This characterizes a distorted hexagonal dense packing. It would be ideal if the value of the x parameter was 0.25 instead of 0.3053. The octahedral interstices of the structure are occupied by the small Ti^{4+} cations.

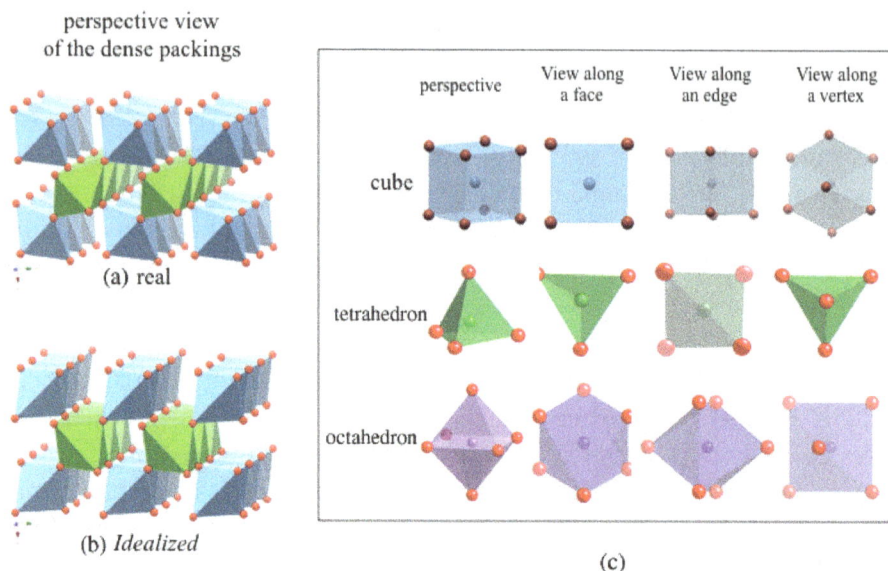

Fig. 3.6.   (a, b) Perspective view of rutile when real, corrugated or ideal dense packings are considered [(a) real when $x = 0.3053$ and (b) idealized when $x = 0.2500$] (c) different perspective and projection views of the most common polyhedra depending on the direction of observation.

As indicated in the legend of Fig. 3.5, oxide ions form an almost compact hexagonal packing, in which the octahedral interstices are occupied by Ti^{4+}. This alternative way [Fig. 3.6(a)] used for the description of the rutile structure will be very useful when the relations which exist between apparently different structures will be studied later (Chapter 6).

This study shows that, depending on the direction of projection, octahedra are seen differently. Fig. 3.6(c) gives some of the principal representations of the most encountered polyhedra.

The crystal chemistry study will be complete when two other pieces of information will be provided: (i) the number of TiO_2 motifs per cell and (ii) the density.

After Chapter 1, the first is now trivial. For titanium, there are 8 ions at the vertices of the cell (therefore $8 \times \frac{1}{8} = 1$ atom) and another one at the center, which gives 2 Ti; for oxygen, 4 are on the faces ($4 \times \frac{1}{2} = 2$) and two inside, which means 4 O; therefore, there are 2 TiO_2 motifs per cell: $Z = 2$. It is worthy to note that, as soon as the significance of the Wyckoff positions is understood, the above calculation becomes unnecessary: Ti being in $2a$ and O in $4e$, there are *a fortiori* 2 TiO_2 motifs per cell. The second piece of information derives from the already determined formula giving the density: $\rho = Z \cdot MM/\mathcal{N} \cdot V$ [MM: molar mass (here 79.9 g.mol^{-1}); $\mathcal{N}$: Avogadro number; V: cell volume]. For TiO_2, ρ is 4.25 g $\cdot$ cm^{-3}.

This is the last step for a standard crystal chemistry study, which completes the initiation of the reader to this discipline. As promised, he/she can verify that the mathematical tools used here are not extremely difficult, and now at his/her disposal are all the tools for a successful description of a structure... This needs to be consolidated now by numerous examples, chosen for the progressive knowledge of the most frequently encountered structure types. They are classified by their chemical formula, according to their decreasing M/X ratio (M metal; X anion). For each structure, just the crystallographic table and the structural results are provided for verification. The reader can then develop the differents steps leading to the final result.

3.2 The Principal Structure Types (Wyckoff, 1968)

3.2.1 *The M_2X type (M/X = 2)*

3.2.1.1 *Li_2O (Fig. 3.7)*

Li_2O			space group F $m - 3m$ (n° 225)
Cubic	$a = 4.873$ Å		$Z = 4$
Li 8c	¼ ¼ ¼; ¼ ¼ ¾	+lattice translations F (½ ½ 0; ½ 0 ½; 0 ½ ½)	
O 4a	0 0 0	+lattice translations F (½ ½ 0; ½ 0 ½; 0 ½ ½)	

This arrangement is similar to another which has been already described: the CaF_2 fluorite structure, but this time, the positions of cations and anions are inverted: the oxide anion takes the place of the cation Ca, and lithium the place of fluorine. Li_2O is obviously called the **antifluorite** structure.

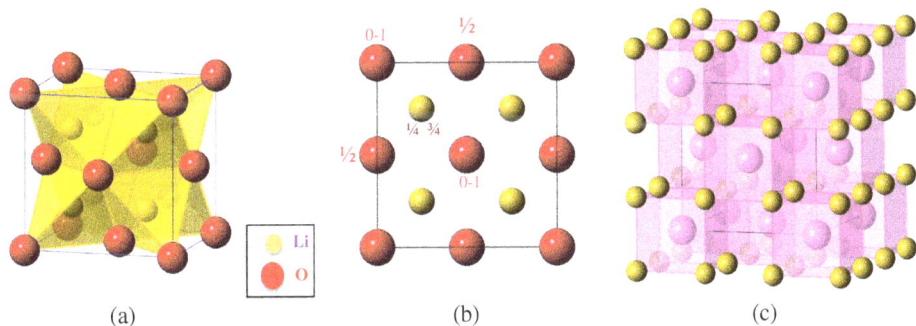

Fig. 3.7. (a) Perspective view of Li_2O; it proves the tetrahedral coordination of lithium in the structure; (b) projection of Li_2O on the square basis; (c) perspective view of Li_2O exhibiting the cubic coordination of oxygen atoms in the structure; (d_{Li-O} = 2.110 Å; ionic bond).

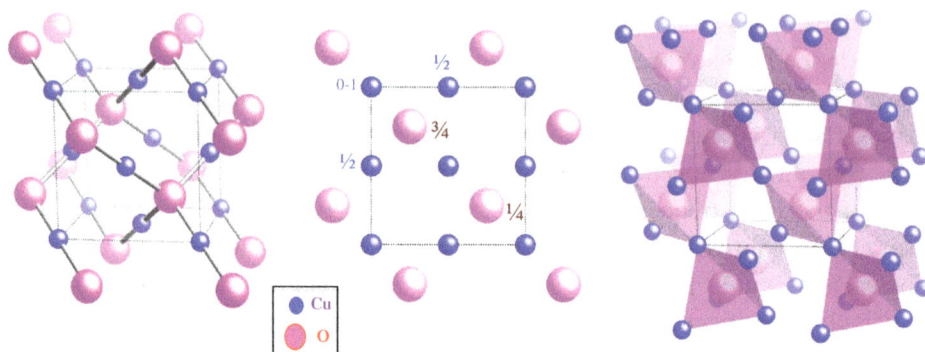

Fig. 3.8.   (a) Perspective view of Cu_2O; it proves the two-fold coordination of copper(I) in the structure; (b) projection of Cu_2O on the square basis; (c) perspective view of Cu_2O exhibiting the tetrahedral coordination of oxygen atoms in the structure; (d_{Cu-O} = 1.848 Å; ionic bond).

3.2.1.2 *Cu_2O (Fig. 3.8)*

Cu_2O	space group P n – 3 (n° 201)	
Cubic	a = 4.267 Å	Z = 2
Cu 4b	0 0 0; ½ ½ 0;	½ 0 ½; 0 ½ ½
O 2a	± (¼ ¼ ¼)	

This two-fold coordination is a characteristic of copper(I). It is often called: "dumbbell" coordination.

3.2.2 *The MX type (M/X = 1)*

The structures of CsCl (*Cs in VIII cubic coordination*) and NaCl (*Na and Cl in octahedral coordination*), already described, will not be recalled. This paragraph presents three other MX structure types: nickel arsenide NiAs and the two allotropic varieties of zinc sulfide ZnS.

3.2.2.1 *NiAs (Fig. 3.9)*

NiAs	space group P 6_3mc (n° 186)		
Hexagonal	a = 3.6 Å	c = 5.01 Å	Z = 2
As 2a	0 0 0; 0 0 ½		
Ni 2b	± (⅓ ⅔ ¼)		

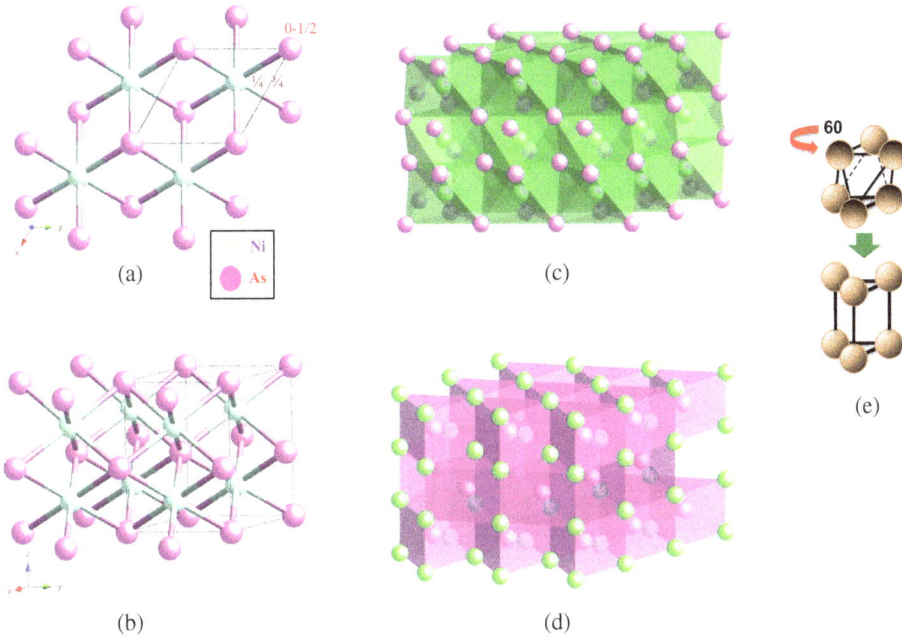

Fig. 3.9. (a) Projection of NiAs on the hexagonal basis; (a) perspective view of atoms and bonds of NiAs; it proves the octahedral coordination of nickel in the structure; (c) perspective view of NiAs exhibiting the VI octahedral coordination of Ni^{2+} in the structure; (d) perspective view of NiAs exhibiting the VI prismatic coordination of arsenides in the structure; (d_{Ni-As}: 2.428 Å; metallic bond); (e) transformation of an octahedron into a triangular prism by a 60° rotation of the upper triangle.

The stacking of arsenides is hexagonal compact [Fig. 3.9(b)]. NiAs provides another example in which the coordination polyhedra are different. The environment of nickel is octahedral [Fig. 3.9(a)]. In the hexagonal base, the nickel octahedra share edges, and form linear columns joined by edges, in the two directions **a** and **b**. The compact octahedral sheets are stacked along the **c** axis by sharing the time faces from one plane to the other. On the contrary, the arsenide environment is trigonal prismatic [Fig. 3.9(c)], but the connections within the (**a, b**) planes and between two sheets along **c** are the same as for nickel.

3.2.2.2 *Wurtzite: The hexagonal variety of ZnS (Fig. 3.10)*

ZnS wurtzite form	space group P 6_3mc (n° 186)		
Hexagonal	$a = 3.81$ Å	$c = 6.23$ Å	$Z = 2$
Zn 2b	($\frac{1}{3}$ $\frac{2}{3}$ z)	($\frac{2}{3}$ $\frac{1}{3}$ z + $\frac{1}{2}$) with z = 0	
S 2b	($\frac{1}{3}$ $\frac{2}{3}$ z)	($\frac{2}{3}$ $\frac{1}{3}$ z + $\frac{1}{2}$) with z = $\frac{3}{8}$	

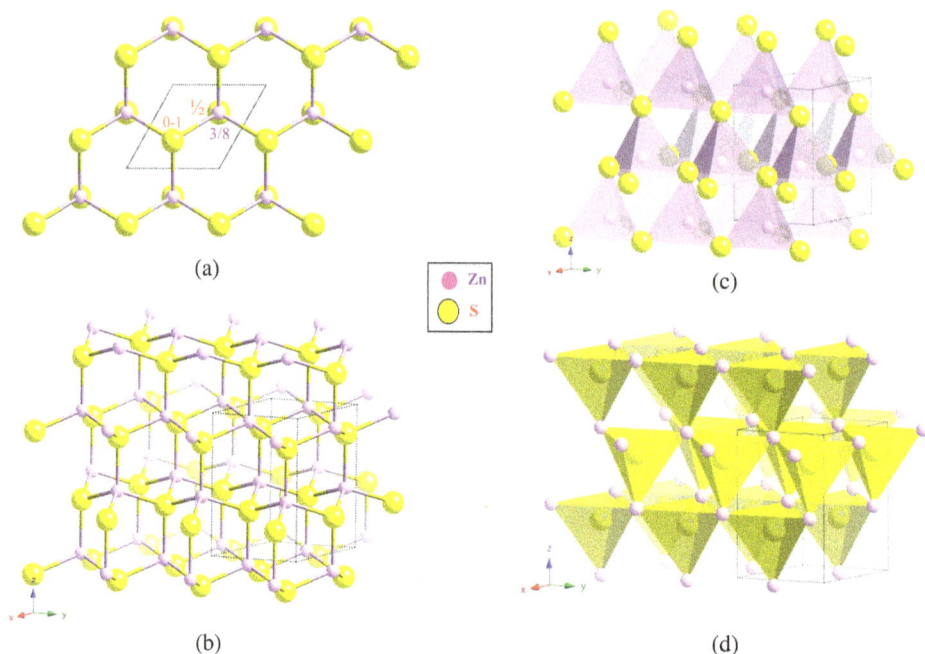

Fig. 3.10. (a) Projection of ZnS on the hexagonal basis; Zn and S atoms are projected at the same place, but at different heights; (b) perspective view of atoms and bonds of hexagonal ZnS; it proves the tetrahedral coordination of zinc in the structure; (c) perspective view of hexagonal ZnS as an assembly of corner-shared tetrahedra; (d) perspective view of ZnS exhibiting the IV tetrahedral coordination of sulfides in the structure; (d_{Zn-S}: 2.333 Å; ionic bond).

Zn^{2+} ions are located on one-half of the interstices of the hexagonal stacking (graphite type) and create a three-dimensional lattice of Zn tetrahedra, sharing vertices.

3.2.2.3 *Blende: The cubic variety of ZnS (Fig. 3.11)*

ZnS blende form		space group F–43m (n° 216)
Cubic	a = 4.873 Å	Z = 4
Zn 4c	¼ ¼ ¼	+ lattice translations F (½ ½ 0; ½ 0 ½; 0 ½ ½)
S 4a	0 0 0	+ lattice translations F (½ ½ 0; ½ 0 ½; 0 ½ ½)

It is interesting to compare the two structures of blende ZnS and fluorite CaF_2 (already described) in terms of occupation of the tetrahedral sites. In fluorite, the fluoride anion occupies all the tetrahedral intersticial sites; in blende, only one-half is filled in an ordered way by Zn^{2+} ions.

(a)

Zn
S

(c)

(b)

(d)

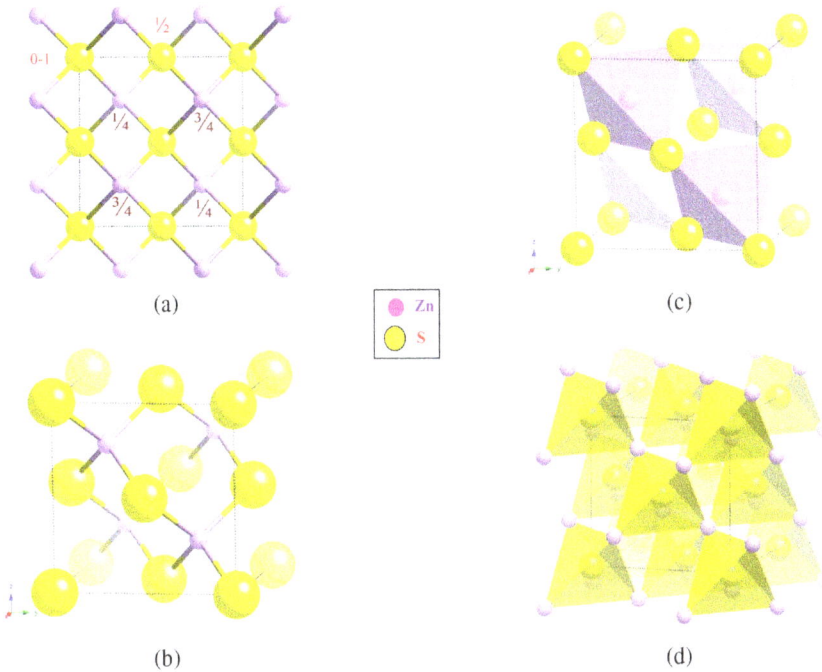

Fig. 3.11.   (a,b) Perspective and projection views of face centered cubic ZnS; (a) Zn and (c) S are in tetrahedral coordination in the structure (d_{Zn-S}: 2.333 Å; ionic bond).

At this stage of the book, the two varieties of ZnS were treated as two independent structure types despite an identical formulae. They are not independent at all and the structural relations existing between them will be described later, in Chapter 6.

3.2.3 *The M_2X_3 type (M/X = 0.66)*

Two well known structure types correspond to this formulation: the α-form of aluminum oxide (corundum) and the β form of gallium oxide Ga_2O_3.

3.2.3.1 *α-Al_2O_3 form (corundum) (Fig. 3.12)*

α-Al_2O_3 form (corundum)	space group R–3c (n° 167) hexagonal setting		
Rhombohedral	$a = 4.7617$ Å	$c = 12.9947$ Å	$Z = 6$
Al 12c	$\pm (x\,0\,0\,z; 0\,0\,\frac{1}{2} - z)$	+ lattice translations R hex. $\pm (\frac{2}{3}\,\frac{1}{3}\,\frac{1}{3})$	
S 18e	$\pm (x\,0\,\frac{1}{4}; 0\,x\,\frac{1}{4}; -x - x\,\frac{1}{4})$	+ lattice translations R hex. $\pm (\frac{2}{3}\,\frac{1}{3}\,\frac{1}{3})$	

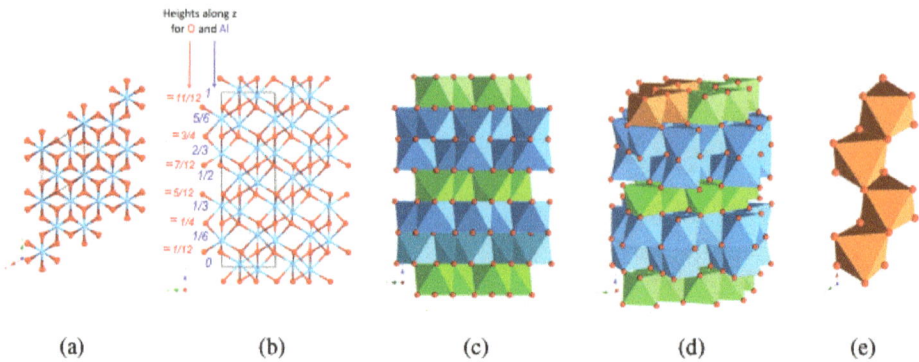

Heights along z
for O and Al

(a) (b) (c) (d) (e)

Fig. 3.12. (a) (001) projection of hexagonal α-Al$_2$O$_3$ (Al: blue), showing the hexagonal dense packing of oxide ions (in red); (b) projection view along an axis perpendicular to the z axis of α-Al$_2$O$_3$; the heights of the different sequences are indicated in blue for aluminium, and in red for oxygen; (c) the same projection using this time polyhedra (here octahedra); for a more comprehensive view, different colors can be used, distinguishing the periodic alternation of the single layers (in green) within which octahedra share edges; (d) the modified view of (c), highlighting the zig-zag octahedral chains (in orange), the repetition of which provides the single layer by edge-sharing; (e) the zig-zag isolated chain (d_{Al-O}: 1.855 Å and 1.972 Å; almost ionic bond). This zig-zag chains will be encountered once more in another form of TiO$_2$, different from the already described rutile.

This structure, seems rather complex at first glance. It provides however the first example in this book where it is necessary to draw several projections, along the three crystallographic axes, for understanding the structure. This method will be often used in the following.

This example illustrates the help of multiple projections in solving the increasing difficulty of deciphering structures. The crystal chemist has two tasks: (i) look very carefully for the details and (ii) consider after detaching from the situation to have a general view of the arrangement. It is like a puzzle: examination of the shape of each piece before assembling them for obtaining the full picture. It is already necessary, but it will be more essential when studying later another variety of aluminum oxide (γ-Al$_2$O$_3$), very complicated, adopting the very different spinel structure.

3.2.3.2 β-Ga$_2$O$_3$ (Fig. 3.13)

β-Ga$_2$O$_3$					space group C 2/m (n° 12)	
Monoclinic	$a = 4.7617$ Å	$b = 3.04$ Å	$c = 5.80$ Å		$b = 103°7$	$Z = 4$
Ga$_1$ 4i	$\pm (x\,0\,z)$	$x = 0.0904$	$z = 0.7948$	+ lattice translation C:		½ ½ 0
Ga$_2$ 4i	$\pm (x\,0\,z)$	$x = 0.3414$	$z = 0.6857$	+ lattice translation C:		½ ½ 0
O$_1$ 4i	$\pm (x\,0\,z)$	$x = 0.1674$	$z = 0.1011$	+ lattice translation C:		½ ½ 0
O$_2$ 4i	$\pm (x\,0\,z)$	$x = 0.4957$	$z = 0.2553$	+ lattice translation C:		½ ½ 0
O$_3$ 4i	$\pm (x\,0\,z)$	$x = 0.8279$	$z = 0.4365$	+ lattice translation C:		½ ½ 0

Its α-form is isotypic with α-Al_2O_3 described above. The table with the β-form shows that all the atoms have heights at only 0 and $\frac{1}{2}$, which renders the lecture of the [010] projection easier [Fig. 3.13(a)]. This structure is very interesting when compared to the α-Al_2O_3 type: despite similar chemical formula (only the nature of the cation changes, with Ga replacing Al, of smaller size), the stacking of anions is different: *hc* for α-Al_2O_3 and *fcc* for β-Ga_2O_3. In the latter, cations are located in the interstices of a *fcc* dense packing, and therefore present two types of coordination: octahedral (in orange) and tetrahedral (in green) [Fig. 3.13(b)]. By sharing edges, the octahedra form double chains, however different from those existing in α-Al_2O_3. Isolated in one layer, their connection is ensured through single chains of corner-sharing tetrahedra which develop in the same direction [Figs. 3.13(c) and 3.13(d)]. Double and single chains are linked by corners [Fig. 3.13(e)].

A final remark, underlying once more that the distance between the cation and the anion is strongly related to the values of their radii as a function of their coordination (d_{Ga-O}: 2.00 Å in octahedra and d_{Ga-O}: 1.83 Å in tetrahedra).

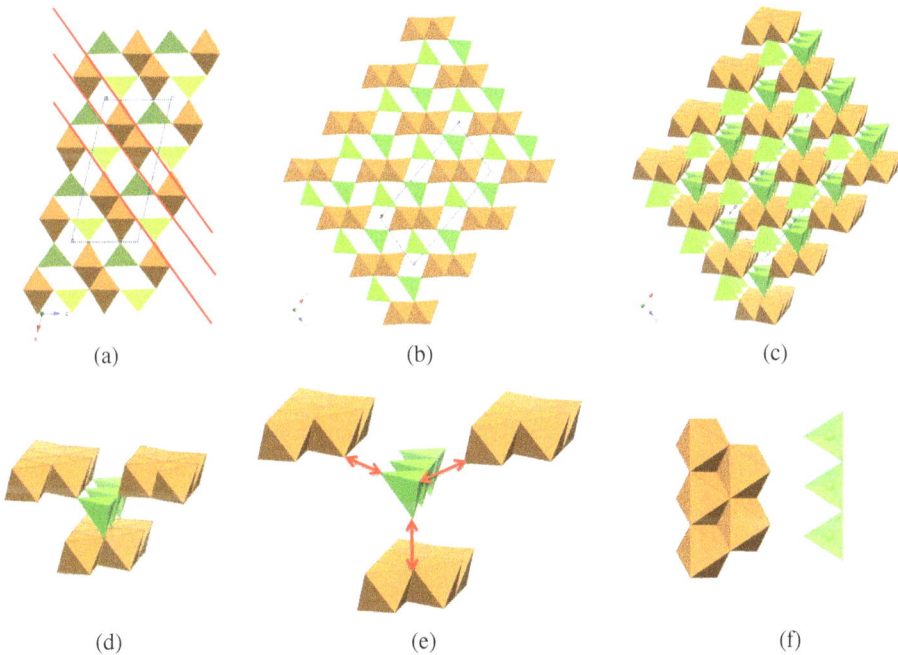

(a)   (b)   (c)

(d)   (e)   (f)

Fig. 3.13. (a) [010] projection of monoclinic β-Ga_2O_3, differentiating Ga^{3+} ions in octahedral (orange, d_{Ga-O}: 2.00 Å) and tetrahedral (green, d_{Ga-O}: 1.83 Å) coordinations. The red lines indicate the direction of oxygen planes; (b) the same projection, this time extended, with a rotation around *y*, for rendering these O planes horizontal; (c) the perspective view proves the edge-sharing octahedral double chains and the single chains of corner-sharing tetrahedra; (d) zoom on the type of junction; (e) dissociated model illustrating the connection by vertices of the double and single chains, which are projected on (f).

3.2.4 *The MX$_2$ type (M/X = 0.50)*

The now well known rutile structure of titanium dioxide will not be recalled here. No mention here of another form of TiO_2, obtained under pressure (HP–TiO_2). It will be presented in Chapter 6 of this book, dedicated to the structural relations between polymorphs.

The structure types described now are both very simple and frequently encountered. Their aristotypes are cadmium halides CdX_2 (X = Cl and I). They also exist for many oxides and sulfides. Being close to each other, they will not be treated separately but just compared.

3.2.4.1 *CdCl$_2$ and CdI$_2$ (Fig. 3.14)*

CdCl$_2$		space group R –3m (n° 160) hexagonal setting	
Rhombohedral	$a = 3.850$ Å $c = 17.460$ Å		$Z = 3$
Cd 3a	0 0 0		+ lattice translations R hex. ± (⅔ ⅓ ⅓)
Cl 6c	± (0 0 z)	$z = 0.2519$	+ lattice translations R hex. ± (⅔ ⅓ ⅓)

CdI$_2$	space group P –3m1 (n° 156)		
Trigonal	$a = 4.24$ Å	$c = 6.840$ Å	$Z = 1$
Cd 1a	0 0 0		
I 2d	± (⅔ ⅓ z)	$z = ¼$	

Two essential characteristics:

(i) The stackings of their triangular planes of anions are both three-dimensional but different: *fcc* for CdCl$_2$, and *hc* in CdI$_2$. When looking at the succession of planes along the axis c of both structures, the octahedral interstices between two planes of anions are alternatively completely filled and form layers or completely empty. This originally empty space is called *Van der Waals gap* (Fig. 3.14) but, by a suitable choice of the charges of cations, it can be partially filled (example: Li$_x$CoO$_2$).

(ii) A precision of vocabulary: the layer structures like those presented here are described as *topologically* two-dimensional, but, in reality, the framework is three-dimensional since the planes are in contact.

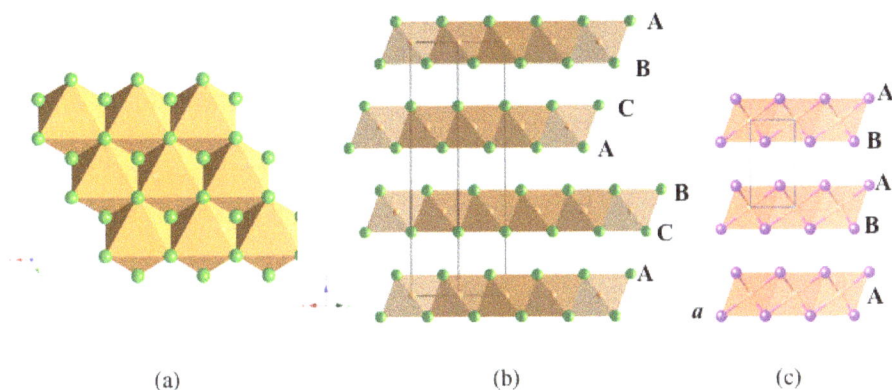

(a)   (b)   (c)

Fig. 3.14. (a) (001) projection of the structures of $CdCl_2$ and CdI_2. Each anionic plane is triangular; (b) projection perpendicular to the z axis of a succession of filled layers in $CdCl_2$; the fcc stacking concerns both the anions (italic lower-case letters), and also the layers of octahedra (bold upper-case letters); (c) the same projection for the hexagonal dense packing of CdI_2 (d_{Cd-Cl}: 2.656 Å and d_{Cd-I}: 3.86 Å; mostly ionic bonds).

3.2.5 The M_3X_4, A_2MX_4 and AM_2X_4 types ($0.75 \geq M/X \geq 0.25$ if $A \neq M$)

The aristotype is the **spinel structure**, one of the oldest known structures (magnetite Fe_3O_4, the first known magnet is a typical example). During the fifties of the last century, this structure was at the origin of numerous magnetic solids, which remain over the years among the best magnetic materials ever discovered. They were also among the best solids for developing examples for different theories of magnetism (Louis Néel, Nobel Prize of Physics 1970).

In spite of its celebrity, it is an extremely complex structure which needs careful examination and many projections for being deciphered, because many atoms of different types are superposed on a single projection.

3.2.5.1 $MgAl_2O_4$ (Fig. 3.15)

$MgAl_2O_4$ direct spinel form	space group F d – 3m (n° 227)	
Cubic	$a = 8.0800$ Å	$Z = 8$
Mg 8a	0 0 0; ¼ ¼ ¼	+ lattice translations F (½ ½ 0; ½ 0 ½; 0 ½ ½)
Al 16d	⅝ ⅝ ⅝; ⅝ ⅞ ⅞; ⅞ ⅝ ⅞; ⅞ ⅞ ⅝;	+ lattice translations F (½ ½ 0; ½ 0 ½; 0 ½ ½)
O 32e	± (x x x; x ¼ – x ¼ – x; ¼ – x x ¼ – x; ¼ – x ¼ – x x) x = ⅜	+ lattice translations F (½ ½ 0; ½ 0 ½; 0 ½ ½)

Even sophisticated programs of drawings are not able to understand this structure, whereas the thoughts of a crystal chemist *a posteriori* help its understanding. The first step consists in a careful examination of the heights of the different atoms. They are all multiples of $\frac{1}{8}$. The complexity of the structure may cause a complete projection to be unreadable. It is better to decompose the projection into fragments and look first at the positions of the cations. Figure 15(a) shows that, for a given height, the aluminum ions Al^{3+} form files at 45° from the cell axes; when the height is changed from z to $\frac{1}{4} + z$ along z, similar files are perpendicular to those at the height z. This observation simplifies the following steps.

Figure 15(b) limits the observation to the fraction of space limited to $0 \leq x \leq \frac{1}{2}$, $0 \leq y \leq \frac{1}{2}$ and $\frac{1}{2} \leq z \leq 1$. The concerned Al^{3+} ions reside at heights $\frac{5}{8}$ and $\frac{7}{8}$ along z. Placing oxygen atoms around them let appear the regular octahedral coordination of aluminium [Fig. 15(b)]. Moreover, considering the two heights $\frac{5}{8}$ and

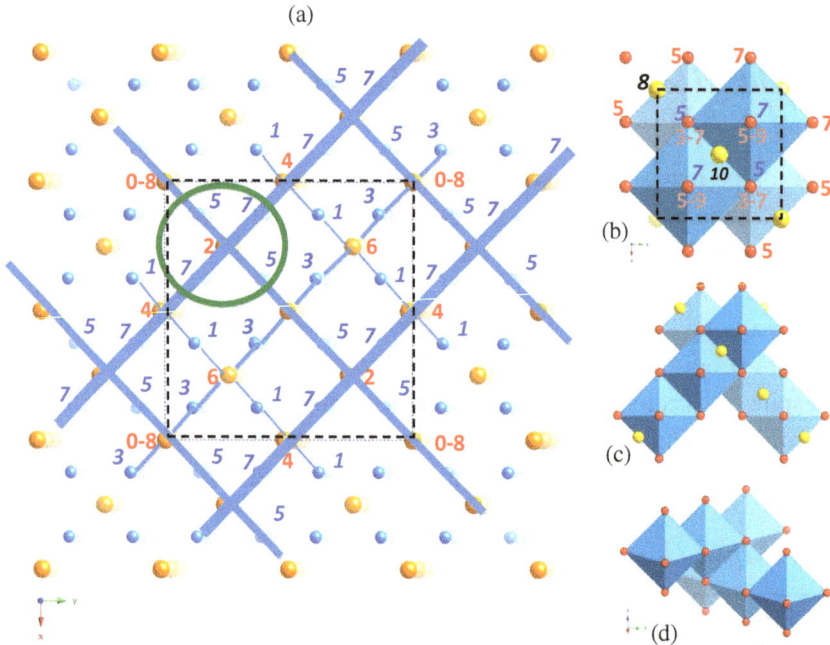

(a)

(b)

(c)

(d)

Fig. 3.15. (a) (001) projection of aluminum (atoms and heights in blue) and magnesium (atoms and heights in orange) in the $MgAl_2O_4$ structure; heights are indicated as multiples of $\frac{1}{8}$ along z; the thicknesses of the lines joining Al files at the same height increase from $\frac{1}{8}$ to $\frac{7}{8}$; (b) zoom on the part inside the green circle of (a) around the heights $\frac{5}{8}$ and $\frac{7}{8}$ for Al^{3+}; oxygen atoms appear in red. The advantage of this view is to prove the regular octahedral coordination of Al^{3+} and their edge-sharing files at the same height; moreover, from the examination of the heights $\frac{5}{8}$ and $\frac{7}{8}$, two files are perpendicular, as shown in the extended projection of (d) and the corresponding perspective view of (e).

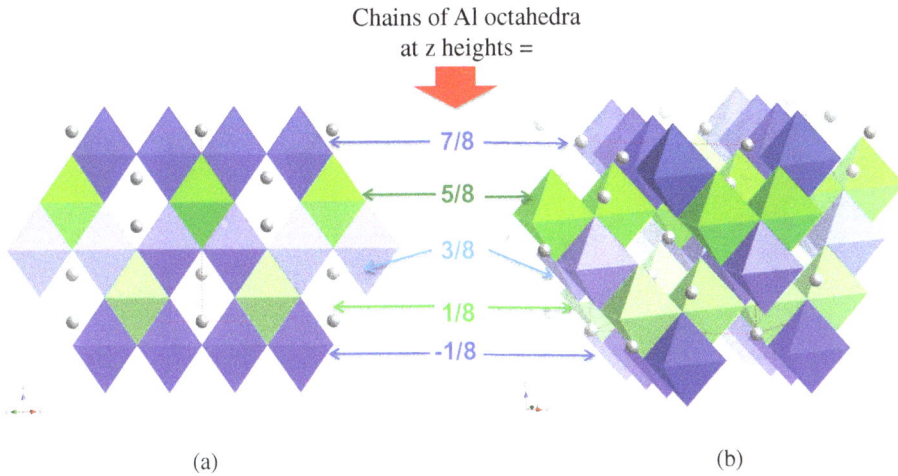

Chains of Al octahedra
at z heights =

(a)   (b)

Fig. 3.16. Arrangement of the orthogonal chains of aluminum in the $MgAl_2O_4$ structure; when blue, the aluminum octahedral chains are at heights $-\frac{1}{8}$, $\frac{3}{8}$ and $\frac{7}{8}$, while those in green are at heights $\frac{1}{8}$ and $\frac{5}{8}$; (a) (-110) projection z; (b) perspective view. The heights along z of Al ions are indicated. Mg^{2+} are grey circles.

$\frac{7}{8}$ for aluminum ions and their oxygenated environment, this figure shows the existence of linear edge-sharing files at both heights but in perpendicular directions. Two files at different levels also share edges. This is clearly seen in Figs. 3.15(c) and 3.15(d). This vision is confirmed for the perspective extension of Fig. 3.16. It proves that the aluminum–oxygen subnetwork can be described by an intersecting tangling up of linear octahedral chains, sharing both their edges within the chains and between them (Fig. 3.16).

In Fig. 3.16(a), the Mg^{2+} ions reside in lozenge tunnels but where are they located in the 3D space? The answer is in Fig. 3.17. They occupy the centers of isolated tetrahedra which share their vertices with the layers containing only octahedra. These layers ((111) planes) are perpendicular to the diagonal of the cubic cell [Fig. 3.17(b)].

From Fig. 3.17(b), two different types of layers, in strict alternation, immediately appear. One of them is built only from aluminium octahedra; the second is an ordered mixture of Al^{3+} octahedra and Mg^{2+} tetrahedra, the latter pointing alternatively up and down. Within this mixed layer, all the connections between the various polyhedra exclusively occur through vertex-sharing. Figure 3.18 gives the details on the nature of each layer and on the connection modes of these polyhedra. In the octahedra–tetrahedra mixed layers, note that one octahedron is surrounded by six tetrahedra alternatively up and down, whereas a tetrahedron has only three octahedral neighbors.

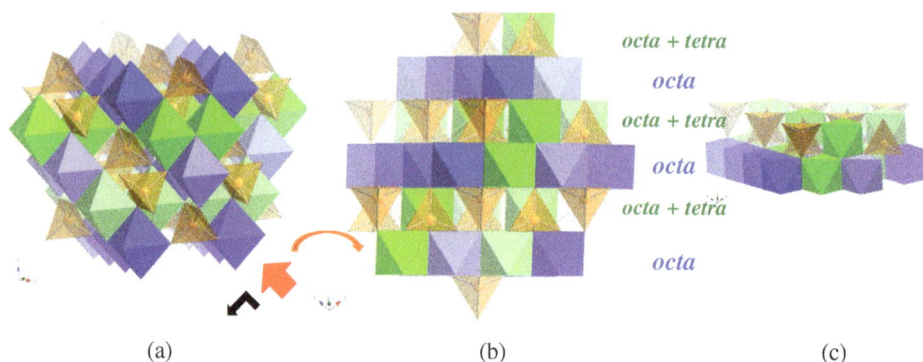

octa + tetra

octa

octa + tetra

octa

octa + tetra

octa

(a)   (b)   (c)

Fig. 3.17. (a) Arrangement of the orthogonal chains of aluminum in the $MgAl_2O_4$ structure, once completed by the Mg^{2+} tetrahedra inserted between the chains; (b) after a rotation of (a), in order to place the planes horizontally, two alternated types of layers appear: those built from only octahedra and those which exhibit both Al octahedra and Mg^{2+} tetrahedra; (c) perspective view of the two different layers and their connection.

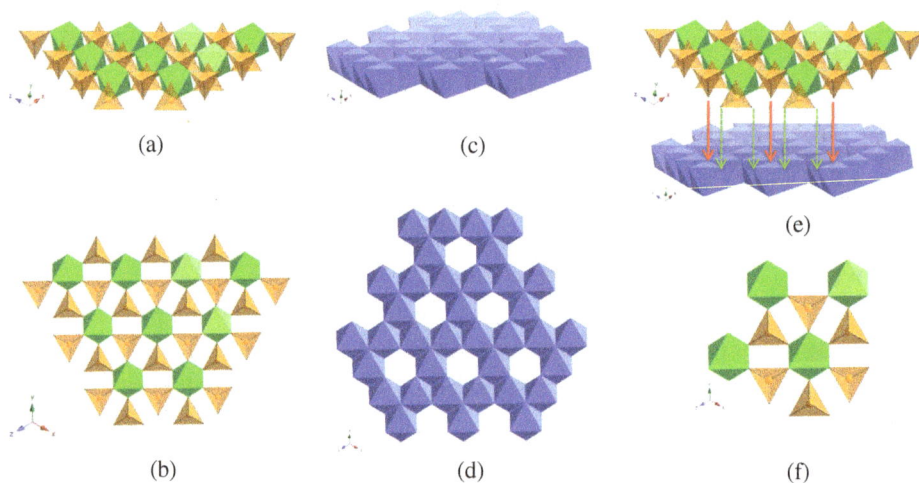

(a)   (c)

(e)

(b)   (d)   (f)

Fig. 3.18. (a, b) Perspective view and (111) projection of the mixed layers, composed of Al octahedra and of Mg tetrahedra; (c, d) perspective view and (111) projection of the layer built exclusively from Al octahedra; such an arrangement is called a Kagomé arrangement; (e) dissociated model illustrating the connection mode between the two types of layers; the red arrows indicate the connection with tetrahedra pointing downwards, and the dotted green arrows those of tetrahedra pointing upwards; (f) partial view of the mixed layer showing both the neighboring of the octahedron surrounded by six tetrahedra, alternatively directed up and down, and the three octahedral neighbors of a tetrahedron.

As already stated, this family is chemically extremely rich because numerous cationic substitutions (complete, partial, ordered or not) can be performed on both the tetrahedral and octahedral subnetworks. This leads to a great variety of physical properties, in particular magnetic. Both the chemical nature of the cations and the different types of cationic ordering lead to extremely different magnetization properties, which lead to magnetic ordering temperatures, which can be tuned in the range 0–700 K. This justifies their use as magnets.

The first magnet is magnetite Fe_3O_4, known since as the Antiquity. Its developed formula: $Fe^{2+}Fe^{3+}_2O_4$ corresponds to two valences for iron. If, in $MgAl_2O_4$, the divalent ion occupied the tetrahedral site and the trivalent the octahedral one, in Fe_3O_4, one-half of the trivalent ions are in the tetrahedra, the other half being in octahedral ones with Fe^{2+} ions. This new situation leads to differentiate the two behaviors: the spinel is called *direct* (case of $MgAl_2O_4$) when octahedral sites are fully occupied by only one kind of atoms. In the other case (magnetite), the spinel is called *inverse*. The respective sizes of the cations play a role in these two cases of distribution.

Well deciphered, this complex structure provides considerable information for the crystal chemist. In such a case, the complexity of a structure is such that, even if sophisticated, no drawing program is able to immediately provide a clear description of the structure. Only a rational approach of a crystal chemist, his step by step progression allows to reach a correct description. After, educating the eye, it is important to look at the structure in all the directions for extracting the different descriptions for the same organization. The simplest is always the one which can be usefully memorized... This being done, the crystal chemist can come back to a rational approach of chemistry. Depending on the structural characteristics he has understood, he can perform pertinent substitutions leading to the creation of new physical properties, transforming the initial solid into a material with interesting performances, which can find innovative applications.

3.2.5.2 $LuFe_2O_4$ (Fig. 3.19)

$LuFe_2O_4$ at 300 K	space group C2/m (n° 12)	
Monoclinic	$a = 5.9563$Å; $b = 3.4372$ Å; $c = 8.6431$ Å; $\beta = 103.24°$	$Z = 2$
Lu 2a	0 0 0	+ lattice translation C ($\frac{1}{2}$ $\frac{1}{2}$ 0)
Fe 4i	$\pm (x\ 0\ z)$ with $x = 0.2084$ and $z = 0.6459$;	+ lattice translation C ($\frac{1}{2}$ $\frac{1}{2}$ 0)
O_1 4i	$\pm (x\ 0\ z)$ with $x = 0.3120$ and $z = 0.8824$;	+ lattice translation C ($\frac{1}{2}$ $\frac{1}{2}$ 0)
O_2 4i	$\pm (x\ 0\ z)$ with $x = 0.1250$ and $z = 0.3843$;	+ lattice translation C ($\frac{1}{2}$ $\frac{1}{2}$ 0)

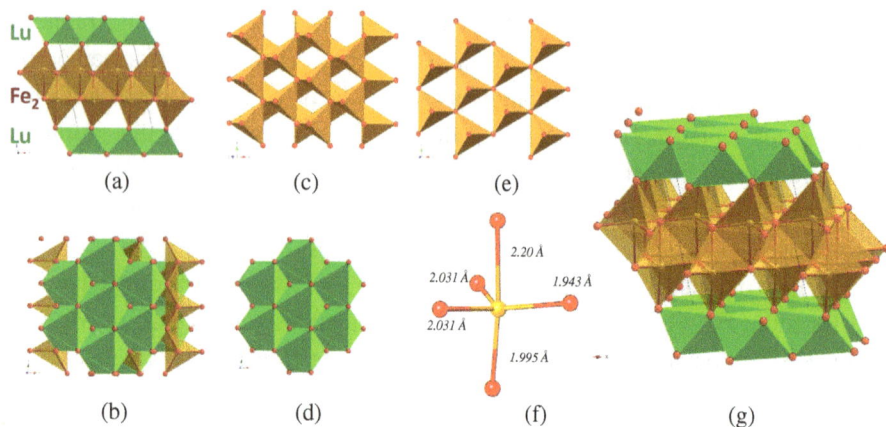

Fig. 3.19. (a–b) the $LuFe_2O_4$ structure projections on (a) the (010) plane and (b) along [010]; (c–d) different [010] projection of the double layer of corner shared iron bipyramids (c) and of the (010) lutetium single layers (d) of edge-shared octahedra; (e) [010] projection of one layer of iron bipyramids, showing their connection by corners; (f) perspective view of the five-fold coordination of iron in $LuFe_2O_4$; the bipyramid is strongly distorted with a distribution of Fe–O distances in the range 1.94–2.21 Å (mean distance: $\langle$Fe–O$\rangle$ = 2.040 Å); (g) perspective view of $LuFe_2O_4$.

It exhibits a periodic stacking along [001] alternating lutetium single layers of the CdI_2 type and iron-based double layers of corner-shared trigonal bipyramids. In them, iron exists in its two valence states Fe^{2+} and Fe^{3+}. The cationic order between Fe^{2+} and Fe^{3+} is very complex. It was recently shown by electron diffraction that, besides this mechanism, the structure can present, as a function of temperature, incommensurate modulations in three dimensions. These modulations result from very complex phenomena which are far beyond the scope of this book.

3.2.5.3 K_2NiF_4 (Fig. 3.20)

K_2NiF_4		space group I 4/*mmm* (n° 139)		
tetragonal	a = 4.013 Å	c = 13.088 Å		Z = 2
K 4e	± (0 0 z)	z = 0.35		+ lattice translation I (½ ½ ½)
Ni 2a	0 0 0;			+ lattice translation I (½ ½ ½)
F_1 4c	0 ½ 0; ½ 0 0;			+ lattice translation I (½ ½ ½)
F_2 4e	± (0 0 z)	z = 0.15		+ lattice translation I (½ ½ ½)

Its structure is very simple, when compared to some of the preceeding ones.

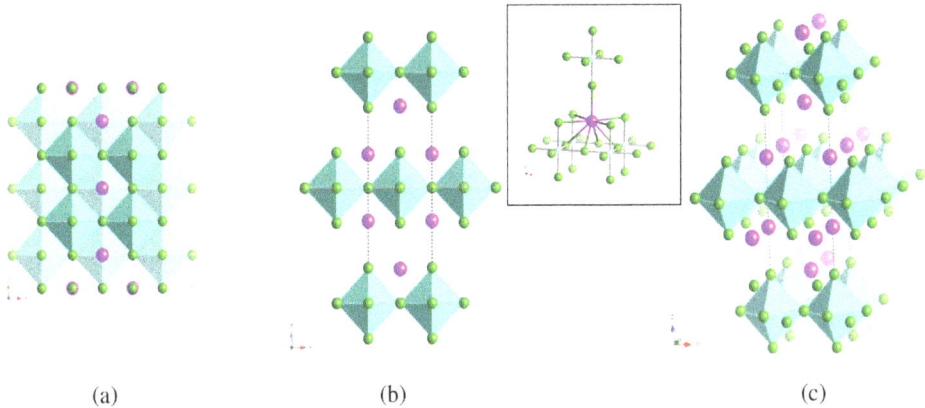

(a)   (b)   (c)

Fig. 3.20.   Projections of the K_2NiF_4 structure (the brighter colors are those close to the eye) on (a) (001) plane and (b) (100) (K^+ in purple, Ni^{2+} are in blue; F^- in green) (c) perspective view of K_2NiF_4; the inset illustrates the nine-fold coordination of potassium (d_{Ni-F}: 2.007 Å and d_{K-F} = 2.624 ($\times$ 1), 2.773 ($\times$ 4) and 2.839 ($\times$ 4) Å; ionic bond).

Nickels are at the center of regular octahedra sharing corners, which form single layers separated by K^+ ions. Two successive planes are shifted by ½a + ½b. This three-dimensional structure is *topologically* described in terms of layers for reasons of convenience. In Chapter 5, devoted to the study of the structural relations between solids, it will be seen that the K_2NiF_4 type is the first term of the series $A_{n+1}B_nX_{3n+1}$ or AX. n ABX$_3$, known as the Ruddlesden–Popper series (from the names of those who, for the first time, had this idea). In this general formula, n characterizes the number of corner-connected single octahedral layers. For example, in the second term of the series: $K_3Ni_2F_7$, the single layers are replaced by double layers of vertex-sharing octahedra, in the interstices of which supplementary K^+ ions are inserted. $n \rightarrow \infty$ corresponds to the three-dimensional perovskite structure described in the next paragraph, devoted to MX$_3$ frameworks.

3.2.6 The MX$_3$ type (M/X = 0.33)

3.2.6.1 ReO$_3$ (Fig. 3.21)

It is probably the simplest structure of the mineral world.

ReO$_3$	space group P m – 3m (n° 221)	
cubic	a = 4.013 Å	Z = 1
Re 1a	0 0 0	
O 3d	½ 0 0; 0 ½ 0; 0 0 ½	

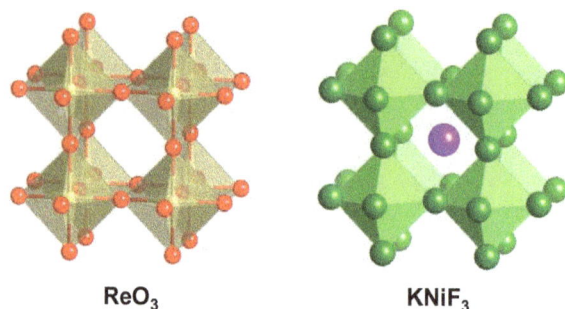

ReO₃   KNiF₃

Fig. 3.21.   Perspective view of two examples of the MX$_3$ framework; in ReO$_3$, the cuboctahedral interstice is empty; in KNiF$_3$, it is occupied by K$^+$, a large cation with small charge.

Its topology was already evoked in Chapter 1 (Fig. 1.14) when, starting from the primitive cubic arrangement, tangent spheres were replaced by octahedra. ReO$_3$, is then a 3D primitive cubic assembly of vertex-shared octahedra. In the center of each cell exists a cuboctahedral interstice, empty in ReO$_3$ but which can be filled by other cations after a related modification of the valence of the cations of the framework (case of KNiF$_3$, Fig. 3.21). The X subnetwork accepts many anions: oxides, fluorides… single or mixed, ordered or not. For example, NbO$_2$F belongs to the ReO$_3$ type, but oxide and fluoride ions are randomly distributed on the X site.

At first glance, this framework seems rigid. It is by no means the case. Mechanically, it allows multiple coordinated movements of the octahedra, in which the X anions play the role of a knee-cap. They will be detailed in the next chapter, even if it was already evoked in Fig. 1.34, illustrating the crystallographic phase transition at 395 K of FeF$_3$. It is only above 395 K that FeF$_3$ exhibits the pure ReO$_3$ cubic structure type; below, it is distorted while keeping the same topology. It is also chemically a very adaptative structure, accepting numerous substitutions (partial or complete) on the A, B and X sites. Hence, for a long time this chemical family is one of the most studied by researchers. Even now, they always find something new for this family, for instance using perovskites as new performant materials for photovoltaics….

An historical digression

A crystal chemist must always think simultaneously about new chemistry. An example? Instead of thinking in terms of completely filled or empty inerstices, one can imagine their partial filling using appropriate syntheses. Historically, it is dur-ing the study of the system x Na$_2$WO$_4$ $\leftrightarrow$ $(1 - x)$WO$_3$ that the phenomenon was observed for the first time in 1949 by Arne Magnéli, a Swedish chemist. The resulting phases, now called the *Magnéli bronzes*, are among the most famous for the solid state chemists community (the term "bronze" refers more to the colors of these phases than to alloys).

Arne Magnéli (1914–1996) was one of the greatest pioneers of structural chemistry of solids. Being a chemist, crystallographer and an electron microscopist, his achievements contributed a lot to the modern concept of extended defects and, more generally, to that of non-stoichiometry. For twenty years (1966–1986), he was also the general secretary of the Nobel committees on physics and chemistry.

It was a conceptual revolution in chemistry. Indeed, in the 19th century, Joseph Proust (1754–1826) established the laws of defined proportions. It stated that, in a chemical formulae, the ratios between the elements forming the molecules are simple fractions. Even if the English scientist John Dalton (1766–1844) supported this theory, the French Claude Berthollet (1748–1822) disagreed, however without proofs. A scientific debate emerged from these opposite factions. A century later, Magnéli, after long arguments, proved that Berthollet was also right. The ratios can also be irrational, and need another formulation for chemical solids (for instance, Na_xWO_3, x being variable). Thanks to Magnéli, the concept of **non-stoichiometry** emerged. Its multiple facets, taught in solid state chemistry courses, are astonishing.

The first advantage of non-stoichiometry is the creation of mixed-valence solids and of new physical properties (mainly electrical and magnetic). Indeed, starting from the neutral skeleton $M^{p+}(X^{n-})_3$ (n^- being the charge of the anion), the introduction of x A^{n+} cations will force the cation M to partially lower its charge for ensuring the electrical neutrality of the considered solid. The new formulae becomes $[(A^{n+})_x(M^{(p-n)+})_x(M^{p+})_{1-x}X_3]$ (*e.g.* the non-stoichiometric potassium tungstate which corresponds to $[(K^+)_{0.4}(W^{5+})_{0.4}(W^{6+})_{0.6}O_3]$). The same metal exists in two different oxidation states (either in ordered or random positions). This allows electron jumps between them, potentially inducing electronic conduction properties, a function of the x value.

The most famous Magnéli bronzes A_xMX_3 (A = alcaline, M mainly V, Ti, W, Mo for oxides) correspond to three structural types, depending on x: cubic ($x \sim 0$), hexagonal ($x \leq \frac{1}{3}$), and tetragonal ($0.4 \leq x \leq 0.6$) (Fig. 3.22).

Historically, the Magnéli bronzes were the first "berthollides" ever discovered and, even if it is not obvious at first glance, their structures present strong structural relations between them, detailed in Chapter 5. For the moment, attention is just focused on the HTB structure (HTB: Hexagonal Tungsten Bronze).

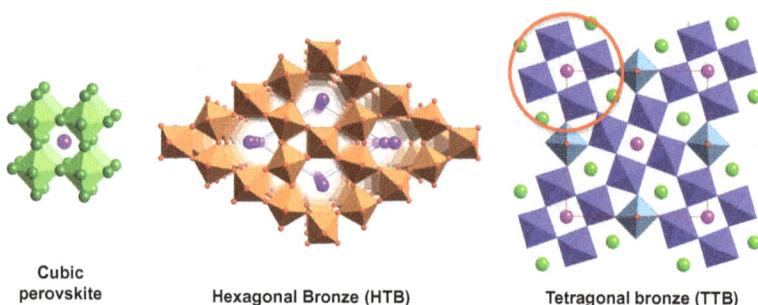

Fig. 3.22.   The most famous Magnéli bronze structures and their usual acronyms (historically, *T* for tungsten). The tunnels are partially or completely filled, depending on *x*. The part of the TTB structure inside the red circle is reminiscent of the perovskite structure.

3.2.6.2 *MX₃ hexagonal framework (Fig. 3.23)*

Similarly with what happened with the chemical insertion of cations within the cuboctahedral interstices of ReO_3, recent "chimie douce" techniques allowed the insertion of cations in the tunnels of the HTB hexagonal phase, first in WO_3, and later in diverse trivalent metal fluorides (M = Al, V, Cr, Fe).

FeF_3 HTB form	space group P 6/*mmm* (n° 191)		
Hexagonal	$a = 7.13$ Å	$c = 3.795$Å	$Z = 3$
Fe 3f	½ 0 0; 0 ½ 0; ½ ½ 0		
F_1 3g	½ 0 ½; 0 ½ ½; ½ ½ ½		
F_2 6l	± (x x 0; x 2x 0; 2x x 0)	x = 0.2114	

Their structures, isotypic with the HTB Magnéli bronze, exhibit two types of channels: large hexagonal, able to host even large cations, and smaller triangular ones, in which insertion is reserved to very small cations.

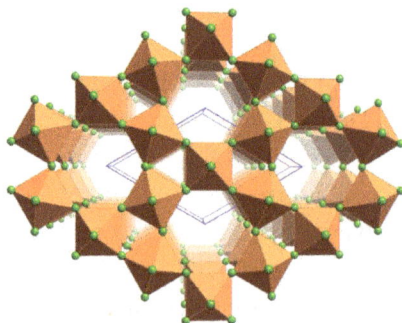

Fig. 3.23.   Perspective view of the hexagonal FeF_3 structure; (iron octahedra in orange, fluoride ions in green).

3.2.6.3 *The MX₃ pyrochlore type with a cubic framework* (*Fig. 3.24*)

Another historical digression

For x close to 0.5, attempts to host large cations (Rb, Cs) with the TTB structure type failed, whereas, with K^+, the non-stoichiometric TTB form was found. Instead, a **stoichiometric** phase, completely different from the TTB structure, appeared. Instead of writing the formulae $A_{0.5}MX_3$, which could imply non-stoichiometry, it must be written $AMM'X_6$ or, in a more detailed mode, $AM^{(p-1)+}M^{p+}X_3$, (A: alcaline ion) for respecting its "daltonian" stoichiometry.

Despite the difference with the TTB structure, its topology was shown to be close to that of mineral pyrochlore (the name pyrochlore comes from the greek *pyros* (fire) and *khloros* (green) according to the green color of the flame when the mineral is put in it). In its natural state, pyrochlore is the oxyhydroxide of a complex mixture of niobium, tantalum and titanium, which hosts also inserted ions alcaline and alcaline-earth ions. Its formula, provided by the mineralogists, is $(Na, Ca)_2(Ti, Nb, Ta)_2O_6(OH, O, F)$. This accounts for the enormous chemical complexity of the solid, as it is very often the case for natural minerals. Within the parentheses of the formulae, the ratios between the species are variable, depending on the geological origin of the studied mineral. However, despite these local discrepencies, the global chemical composition is preserved. It is this global composition which interests the crystallographer or the structural chemist who, by an appropriate and rational choice of the cations and their charges, tries to simplify this complexity. In this search, the mineralogical formulae can be reduced to a general formulation $A_2M_2O_6X$, ignoring in the first step the chemical nature inside each parenthesis.

In the seventies, a lot of attention was paid by solid state chemists to this complex structure. The first relation they found was the relation between pyrochlore and fluorite structures, due to coordinated movements of oxide anions. Later, the French chemist Jean Pannetier proposed a new description of the pyrochlore $A_2M_2O_6X$ consisting of the interpenetration of two different subnetworks AX and AM_2O_6. The pyrochlore structure can exist without the AX subnetwork. This particularity generated two rich independent fields of research: one on $A_2M_2O_6X$ compounds, the other on the defect pyrochlore AM_2O_6 (A: NH_4^+, Rb^+, Cs^+, Tl^+). Both frameworks accept numerous cationic substitutions in the skeleton. In the AM_2O_6 family, the A cation, always of large size, occupies the large cavities existing in the structure. However, during a long period, and at variance to what happened for perovskite and HTB structures whose cages and tunnels can be empty, the cages were always fully occupied in the AM_2O_6 family. The empty pyrochlore was created only in 1989 by the author of the present book and his colleagues. For that, he applied "chimie douce" methods to the brown solid $NH_4^+Fe^{2+}Fe^{3+}F_6$ by oxidizing Fe^{2+} by bromine. This led to the transformation of ammonium into gaseous ammonia which helped to evolve the pyrochlore form of FeF_3 (pale green) that was then discovered.

Despite its simple formulae, the pyrochlore structure is complex and the method already developed for spinel needs to be applied, using structural relations of pyrochlore. The first of them is their identical space group F d – *3m*.

FeF_3 pyrochlore form		space group F d – *3m* (n° 227)*
Cubic	a = 8.0800 Å	Z = 16
Fe 16c	0 0 0; ¾ ¼ ½; ¼ ½ ¾; ½ ¾ ¼	+ lattice translations F (½ ½ 0; ½ 0 ½; 0 ½ ½)
F 48f	± (x ⅛ ⅛) with x = 0.9341	+ other positions given in the International Tables + translations of the F network

*In the International Tables of Crystallography, the space group F d – 3m is described in two ways, that depend on the choice of the origin of the cell. The data presented here correspond to the choice 2 of International Tables.

As for spinel, the (100) projection [Fig. 3.24(a)] and a careful examination of the heights of iron ions show that these ions are organized in files, orthogonal to each other from one height to the other. The only difference at this level is that, this time, octahedral files share corners instead of edges.

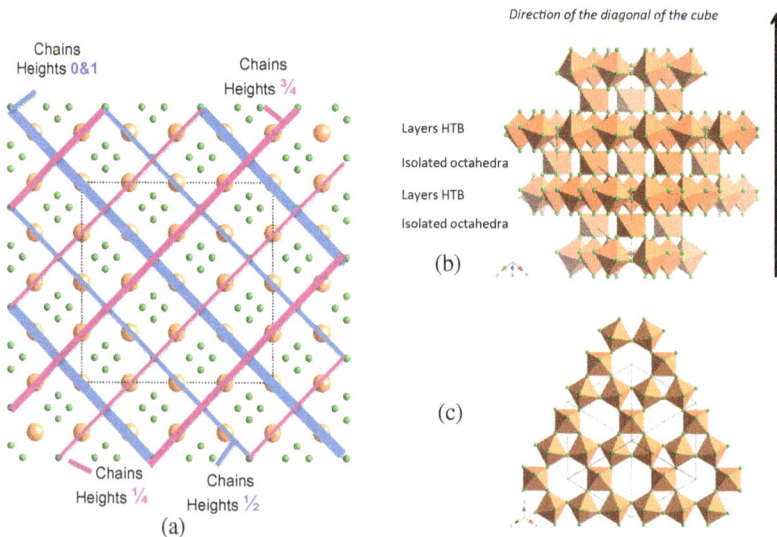

Fig. 3.24.   (a) (100) Projection of the FeF_3 pyrochlore structure (orange spheres: iron; green spheres; fluorine); (b) projection in a direction perpendicular to the diagonal of the cubic cell; as for spinel, the figure appears as a successions of layers, alternatively dense or not; (c) projection of a dense layer, which curiously shows a HTB arrangement.

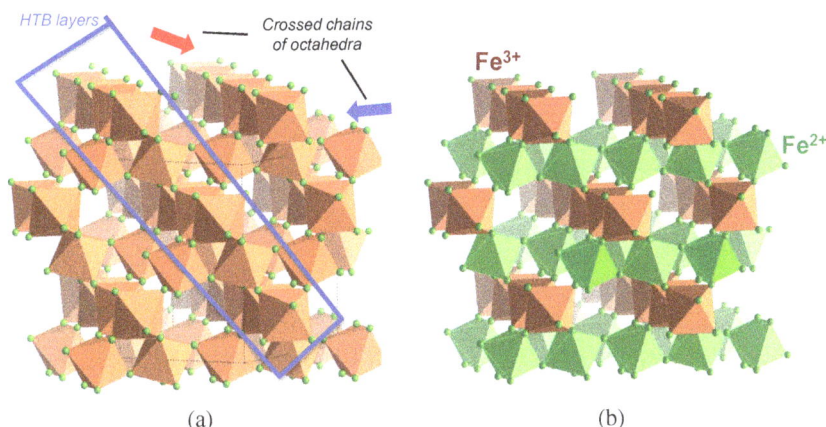

Fig. 3.25. (a) Perspective view of the cell of the pyrochlore form of FeF_3; it represents the two modes of description: orthogonal chains and HTB planes; (b) the same view for $(NH_4^+Fe^{2+}Fe^{3+}F_6)$ the oxidation of which led to the formation of pyrochlore FeF_3; the topology is preserved but with an ordering between Fe^{2+} and Fe^{3+} orthogonal chains.

Looking at the structure in a direction perpendicular to the diagonal of the cubic cell reveals a succession of alternated layers with different densities, but always sharing corners of the iron octahedra. Isolation of the denser layer shows that its topology is exactly the same as that existing in HTB-FeF_3 [Figs. 3.24(b) and 3.24(c)]. This provides another description of the pyrochlore structure: the alternation of HTB layers connected by isolated iron(III) octahedra. The perspective view of Fig. 3.25 gathers the two possible descriptions, both simple but complementary.

The structure of $NH_4Fe^{2+}Fe^{3+}F_6$, the initial product leading by oxidation to the new form of FeF_3, is also represented in Fig. 3.25(b). The strict cationic order existing between Fe^{2+} and Fe^{3+} has two consequences: (i) a differentiation between two successive orthogonal chains, one exclusively occupied by Fe^{3+} and the other by only Fe^{2+}; (ii) from the crystallographic point of view, a lowering of the symmetry: orthorhombic for $NH_4Fe^{2+}Fe^{3+}F_6$, and cubic for FeF_3 with a double volume for the cubic cell. This a good example of a topotactic reaction which maintains the topology constant.

This example induces a remark for a crystal chemist. When he discovers an interesting structure, he can, even at constant topology, change drastically the properties of this structure by using some creative chemistry. The example of three *a priori* different varieties of FeF_3 is particularly illustrative. If, at first glance, there are very few structural relations between them, they are thermodynamically related. Indeed, a progressive heating of the pyrochlore form leads to its transformation first into the HTB form, and later into the cubic form. This gives a good idea of the relative thermodynamic stabilities of these forms. Moreover, in terms of properties, despite being all antiferromagnets, they have very different Néel magnetic ordering temperatures: 365 K for the cubic form, 110 K for HTB and only 20 K for the pyrochlore form. This is only due to the type of adopted structure, the values of the different Fe–F–Fe angles playing a drastic role. Note also that this example was at the origin of a new physical concept: the ordered magnetic frustration.

So crystal chemistry is not always an end; it can also be the beginning for the rational discovery of new innovative materials, a concept which will be developed in the last chapters.

In the above examples, the MX_3 structure types were always three-dimensional, with corner-shared polyhedra. A question arises now: is it always true, or can two-dimensional solids, in which edges are shared, exist with the same MX_3 formula? Such is the case of molybdenum oxide MoO_3.

3.2.6.4 *MoO_3 (Fig. 3.26)*

MoO_3		space group P *bnm* (n° 62)		
orthorhombic	$a = 3.963$ Å	$b = 13.855$ Å	$c = 3.6964$ Å	$Z = 4$
All atoms in $4c$ $[\pm (x\ y\ ¼;\ ½ - x\ ½ + y\ ¼)]$				
Mo $4c$	$x = 0.0867$	$y = 0.1016$		
O_1 $4c$	$x = 0.4494$	$y = 0.4351$		
O_2 $4c$	$x = 0.5212$	$y = 0.0866$		
O_3 $4c$	$x = 0.0373$	$y = 0.2214$		

The projections of Fig. 3.26 clearly show the existence of a stacking along [010] of isolated double layers parallel to the xOz plane of the orthorhombic cell [Fig. 3.26(c)]. Two successive layers are shifted by one-half of the *a* parameter. The structure is built from double chains of octahedra sharing vertices and edges [Figs. 3.26(a) and 3.26(b)]. The molybdenum octahedra are highly distorted with distances in the range $\langle 1.67 - 2.33$ Å$\rangle$ (2.33; 2.25; 1.95 (twice); 1.73Å and 1.67Å). These last two values correspond to the Mo=O double bonds existing in the octahedron.

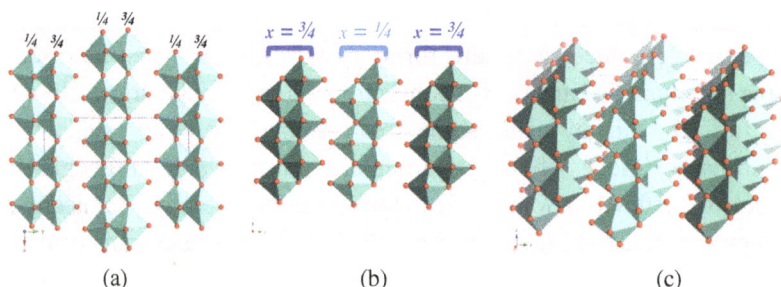

(a)   (b)   (c)

Fig. 3.26.   (a) [001] and (b) [100] projections of the MoO_3 structure (Mo: blue; O: red). (c) The perspective view proves the infinite double layers parallel to the (010) plane.

3.2.6.5 $YBa_2Cu_3O_7$, the guest and its variants

$YBa_2Cu_3O_7$		space group P mmm (n° 47)		
orthorhombic	$a = 3.824$ Å	$b = 3.888$ Å	$c = 11.690$ Å	$Z = 1$
Y 1h	½ ½ ½			
Ba 2t	$\pm$ (½ ½ z)	with $z = 0.1854$		
Cu_1 1a	0 0 0			
Cu_2 2q	$\pm$ (0 0 z)	with $z = 0.3557$		
O_1 1e	0 ½ 0			
O_2 2s	$\pm$ (½ 0 z)	with $z = 0.3775$		
O_3 2q	$\pm$ (0 0 z)	with $z = 0.3786$		
O_4 2q	$\pm$ (0 0 z)	with $z = 0.1582$		

Due to its extraordinary superconductivity properties and the immense international success of this solid, it deserves inclusion in the list of the most famous structures. It is an yttrium–baryum cuprate with formula $YBa_2Cu_3O_7$, often written YBaCuO (from its symbols) or *123* with only the ratio between the cations. It is an **ordered** perovskite structure. The great chemical adaptability of this structure type, one of the most studied in the history of sciences, was already mentioned. YBaCuO provides another extraordinary example, both from the structural and properties points of view.

But, before showing its characteristics, let us consider some historical aspects. This is important because its discovery created a revolution for all solid state scientists.

With its related solid (Sr, La)Cu_2O_4, (K_2NiF_4 type), also a superconductor below 35 K, it completely changed what was seen before for superconductivity. Many researchers (including Nobel laureates!) considered that the domain had reached its maximum possibilities after the discovery of the metallic alloy Nb_3Ge and its superconductive behavior below 23 K... For YBaCuO, it was below 93 K, above the boiling temperature of liquid nitrogen! A revolution: oxides — and not only metals — could be superconductors! Not only the Swiss researchers Georges Bednorz and Alex Müller (IBM Zürich) received the Nobel Prize of Physics in 1987 for this discovery but this led to also an extraordinary international attention from both chemists and physicists of the solid state. Thousands of researchers, worldwide, changed their topic of research to invest in this new domain. By the way, the world record: 138 K belongs to mercury cuprates.

Beyond history, why is YBaCuO so interesting for the crystal chemist? Several reasons can be presented: (i) the oxygen composition is continuously variable, from $YBa_2Cu_3O_6$ to $YBa_2Cu_3O_8$, depending on the applied oxygen pressure. This rare behavior must be structurally explained; (ii) despite the variations of the chemical formulae, the global topology of the structure is maintained, as it was previously the same for ReO$_3$ and perovskite; and (iii) as only the valence of copper changes with the various amounts of oxygen, it is expected that, simultaneously, the coordination polyhedra of copper will also change. This last point will give the opportunity to introduce a new notion during the description of the structure which follows: **the concept of vacancy**.

Look first at the three projections of the structure along the three axes of the orthorhombic cell [Figs. 3.27(a)–3.27(c)]. They provide complementary information. Figure 3.27(a) recalls the projection of the aristotype perovskite structure described in terms of polyhedra (François *et al.* 1988). Figures 3.27(b) and 3.27(c) present the two types of coordination adopted by copper: Cu1 exhibits a square plane environment (SP) whereas Cu2 is surrounded by five oxygen atoms which form a square-based pyramid (SBP). The structure can then be described by the stacking of triple layers of copper polyhedra sharing corners, and parallel to (001). They are separated by yttrium ions. Y and Ba subnetworks within the perovskite cages are also ordered with a strict Ba–Ba–Y alternation. Barium ions [Fig. 3.27(f)], within the perovskite cages, wrap the middle square-plane chains of copper. They adopt a ten-fold coordination (12 in the aristotype perovskite). Yttrium has eight neighbors in a cubic environment [Fig. 3.27(e)].

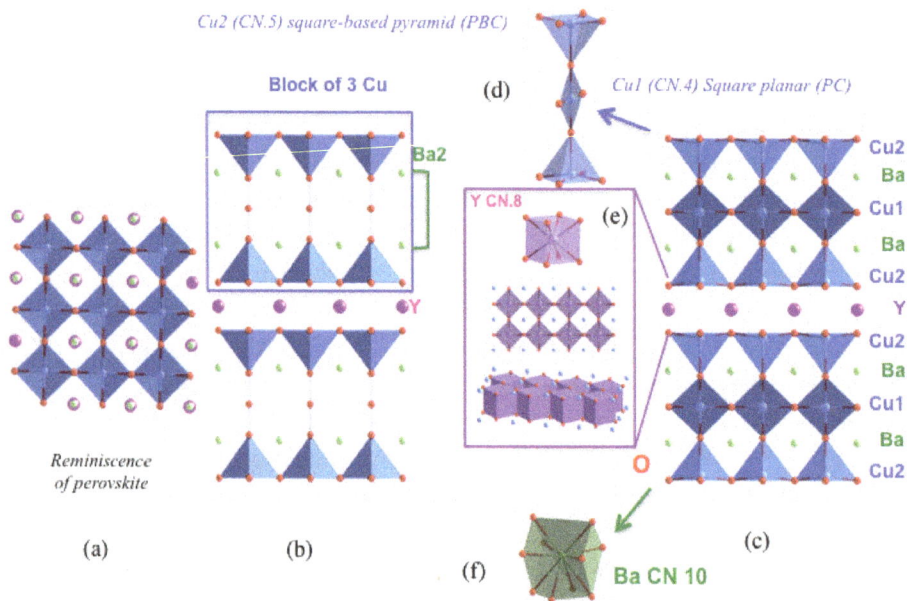

Fig. 3.27.   (a) [100] projection of the aristotype perovskite; (b,c) [100] (b) and [010] (c) projections of YBa$_2$Cu$_3$O$_7$. (d)–(f) highlight the coordinations of the different metals, particularly those of copper: square-plane (SP) for Cu1, square-based pyramid (SBP) for Cu2. The polyhedra are almost regular (distances Cu-O = 1.849 Å and 1.944 Å for Cu1, 1.929, 1.962 and 2.309 Å for Cu2, 2.745, 2.890 and 2.959 Å for Ba–O and 2.414 Å for Y–O).

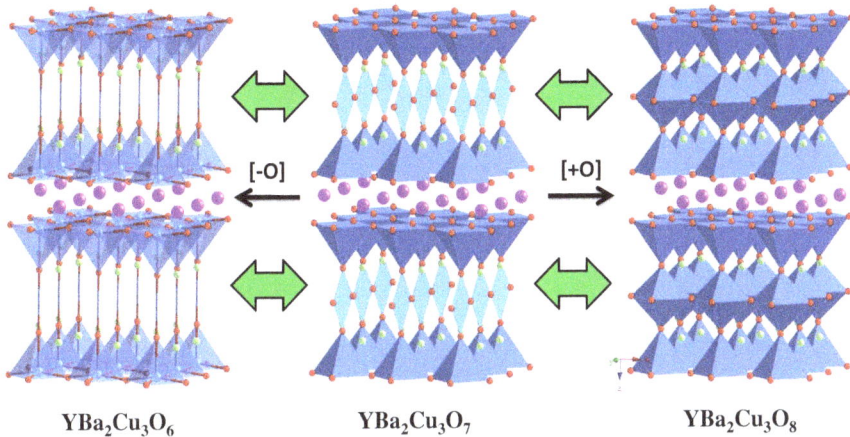

Fig. 3.28. Perspective views of the structural changes occuring when the compositions change from $YBa_2Cu_3O_6$ to $YBa_2Cu_3O_8$ as a function of the oxygen amount in the formula. The Cu1 coordination, originally with two oxygen atoms (like the dumbbell in Cu_2O) becomes 4 in a square plane for $YBa_2Cu_3O_7$, and finally 6 in the octahedra $YBa_2Cu_3O_8$. Correlatively, the barium coordination follows the sequence VIII; X; XII, the latter corresponding to a cuboctahedron, like potassium in the perovskite $KNiF_3$.

It was mentioned above that, during the controlled synthesis, the oxygen amounts in this structure are continuously variable. What is the structural consequence?... Only the Cu1 and Ba environments are affected during this evolution, whereas those of Y and Cu2 remain the same (Fig. 3.28).

Looking at the theoretical oxidation numbers of copper in the three compositions (in reality, the situation is much more complicated!), the composition O_6 corresponds to a couple 1 Cu^+ – 2 Cu^{2+}, O_7 to 1 Cu^{3+} – 2 Cu^{2+}, and O_8 to 3 Cu^{3+}. Even if the reader could be tempted to propose a strict relation between oxidation numbers of copper and their environment, both Cu^+ and Cu^{++} can exhibit several coordination polyhedra, depending on the structure. Moreover, the passage $YBa_2Cu_3O_7 \rightarrow YBa_2Cu_3O_6$ corresponds to the elimination of the O1 site in the table of positions and the coordination goes from IV to II. On the contrary, going from $YBa_2Cu_3O_7$ to $YBa_2Cu_3O_8$, implies in this table to add the position O5 (1b: ½ 0 0) which transforms the square plane into an octahedron. Inversely, starting from $YBa_2Cu_3O_8$, the two others are obtained by suppressing O5 then O1 sites around Cu1. These sites become empty, and they are called **vacancies**.

This concept of vacancies requires a precise definition, explaining the difference between an interstice and a vacancy. An interstice corresponds to an **initially** empty space between atoms. A vacancy is a space **initially occupied** by matter and then later, for external reasons, can disappear; the occupied space becomes

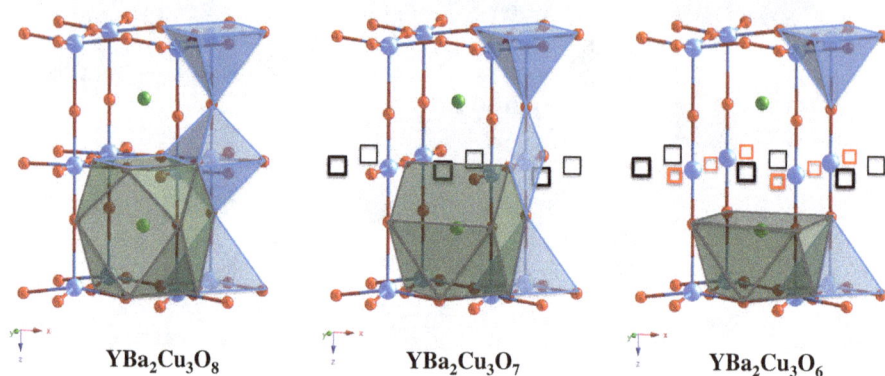

Fig. 3.29. Perspective view of the evolution of the coordination polyhedra of Cu1 and of baryum from oxygen vacancies represented as empty squares (black for O5; red for O1).

empty. Interstices are intrinsic to every structure. Vacancies are often created. They can be either purely cationic or anionic, but both types can exist simultaneously in a structure (Fig. 3.30).

3.3 The Concept of Vacancy; Its Limits with the Existence of Lone Pairs

Initially chemical and structural, these vacancies can also serve as *tools* for proving topological relations between apparently different structures. Two examples on this point: relations between (i) the NbO and NaCl structures [Fig. 3.30(a)] and (ii) the CaF_2 and CsCl ones [Fig. 3.30(b)].

NbO derives from NaCl, with both ¼ of vacant anionic sites (purple squares) and ¼ of vacant cationic sites (green stars). For NaCl, $Z = 4$, which leads to the structural formulae Na_4Cl_4. Therefore, the structural one would be $Nb_3\square_1O_3\square_1$. for NbO ($\square$ represents a vacancy).

In the same way, the fluorite CaF_2 can be viewed as a defective CsCl structure with exclusively cationic vacancies (white squares). The structural formulae of CaF_2 is therefore $Ca_1\square_1F_2$.

Attention must be carefully paid for a right application of the concept of vacancy, which is associated with a really empty site. For some structures, what the eye considers as an empty space using the present conventions for drawings can be wrong, when the space is not occupied by an atom, but by an orbital. It is particularly the case for cations like thallium(I): Tl^+, lead(II): Pb^{2+}, bismuth(III): Bi^{3+} and antimony(III): Sb^{3+}, which possess non-bonding electron pairs (mainly ns^2 lone pairs).

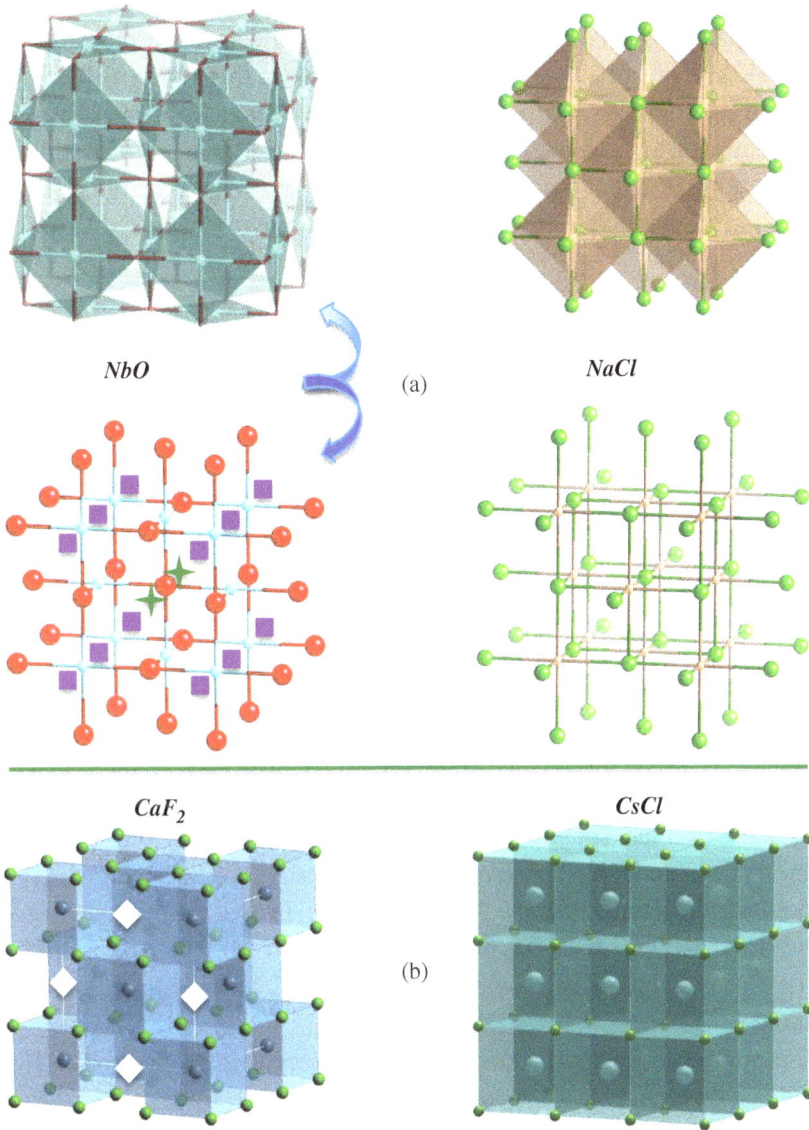

Fig. 3.30. (a) The comparison between NbO and NaCl structures represented in both polyhedral and balls and sticks representations. The structural formula of NbO ($Nb_3\square_1O_3\square_1$) indicates the absence of both ¼ of the anionic sites (purple squares) and ¼ of the cationic sites (green stars); (b) in the same way, compared to that of CsCl, the structural formula of fluorite CaF_2 is $Ca_1\square_1F_2$.

Depending on the structure in which they are involved, the lone pairs ns^2 of these cations occupying a variable volume, which can reach that of the oxide anion (~18 Å^3), are often stereoactive. Non-visible, this lone pair appears as a vacancy in the topology, but is not a vacancy in the sense defined above. Their steric occupation can have structural consequences (changes of multiplicities, symmetry, for example...) and even modify the *apparent* description, leading to a bad interpretation for the crystal chemist who is not aware of this particularity.

A good illustration of this influence is provided by the red form of lead oxide PbO (the other form of PbO is known as the yellow form). Its crystal data are the following

PbO red form	space group P $4/nmm$ (n° 129)		
System: tetragonal	$a = 3.976$Å	$b = 5.023$ Å	$Z = 2$
Pb 2c	$\pm$ (0 ½ z); with $z = 0.237$		
O 2a	0 0 0; ½ ½ 0		

At first glance, the projections [Figs. 3.31(a) and 3.31(b)] indicate a four-fold coordination for Pb with a square plane of oxygen atoms and, apparently, a Van der Waals gap between Pb^{2+} ions, which is rather surprising because of the electrostatic repulsion between positive charges. It could be thought that vacancies (represented by blue squares in Fig. 3.31(c)) exist in the interlayer. If these "vacancies" were occupied by *real* anions, for instance oxides [Fig. 3.31(e)], the Pb coordination would become VIII instead of IV with cubes, and lead to a fluorite structure [Fig. 3.31(g)]. It becomes clear that Pb lone pairs [Fig. 3.31(d)] take the place of anions for providing an apparent 2D aspect [Fig. 3.31(f)] to the structure.

This organization will also be encountered later with Bi(III)-based superconductors.

At the end of this chapter, numerous structures, with different types of polyhedra and connections have been described, giving a small idea of the great richness of the world of solids. But, before ending this chapter, a final comment on the relations between shape and connectivity of polyhedra and the formulae of the corresponding solids.

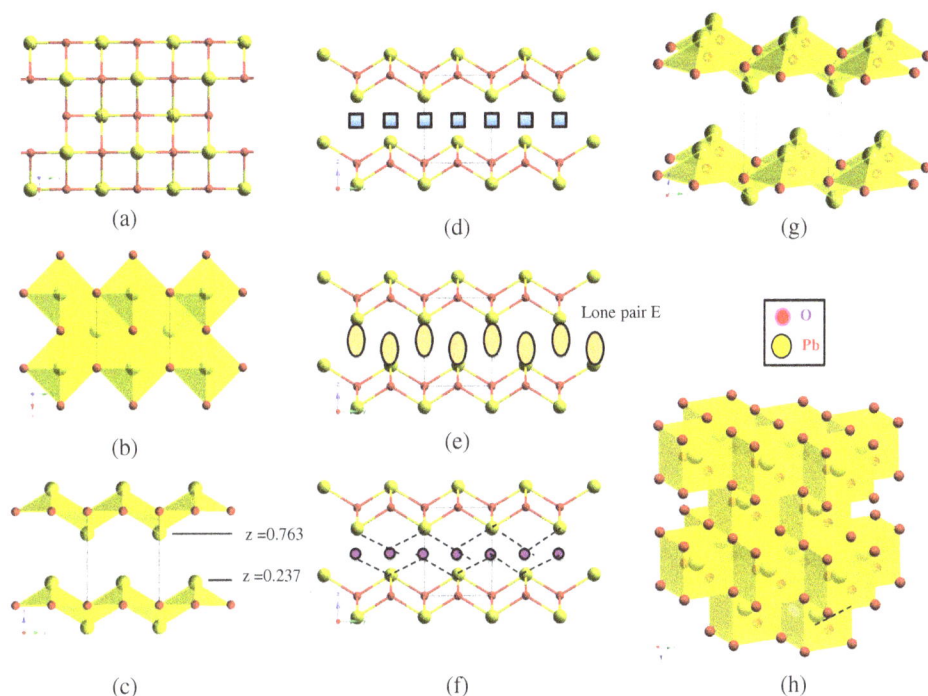

Fig. 3.31. (a,b) [001] and [100] (c) projections of the structure of red PbO; (b) shows the four-fold coordination of lead with its heights; (d) [100] projection considering PbO in terms of vacancies (blue squares) in a Van der Waals gap; (e) same representation with lone pairs (pale orange ovals) instead of vacancies; (f) organization of the structure if the sites were occupied by a real anion (purple) leading (h) to a fluorite structure; (g) perspective view of PbO.

3.4 Polyhedra, Connectivity and Formulation

Implementing an intellectual game, the readers are now able to imagine original associations of polyhedra, like in the Lego games of their childhood. Imagine that they are successful. The structures seem mechanically stable, plausible and the envy emerges: create them chemically! But a problem appears... To what chemical formula (even general) does this assembly correspond? What starting associations of compounds can provide the solution? A good understanding of what was presented in the previous pages can orient them. It will indeed be shown in the following that, knowing their Lego-like intellectual construction, they will be able to arrive at a precise chemical formula in terms of proportions of cations and anions, with their associated polyhedra.

For that, there exist two master-trumps: coordination and connectivity. The coordination will be that of the element $\mathbf{M}$ inside a $\mathbf{MX}_n$ polyhedron. The connectivity concerns the element $\mathbf{X}$ situated at the vertices of the initial p_0 polyhedron. Another term for pointing out the coordination of element $\mathbf{X}$ is also the number of M atoms attached to X, and consequently, the number of polyhedra adjacent to the initial p_0 polyhedron. Using the way followed for the determination of the number of motifs per cell, the $\mathbf{X}$ atom does not belong exclusively to p_0 but also to its $\mathbf{N}$ neighbors. Therefore, only a $1/N$ atom fraction of element $\mathbf{X}$ in p_0.

Take the example of a tetrahedron $\mathbf{MX}_4$ involved in a vertex connection with other tetrahedra (a frequent case for silicates, phosphates and arsenates). It can share 0, 1, 2, 3 or 4 vertices with others (Fig. 3.32). In the first case, the 4 X atoms exclusively belong to the $\mathbf{MX}_4$ polyhedron p_0. They each count for 1 and the formula is $\mathbf{MX}_4$. When p_0 shares one vertex with another polyhedron (whatever its nature), only three X atoms belong to p_0 and count for 1. The fourth, shared with another polyhedron, counts only for ½. Therefore, only $(3 \times 1) + (1 \times ½) = 3.5$ X atoms effectively belong to p_0, the formula of which becomes $\mathbf{MX}_{3.5}$. If the shared polyhedron is also a tetrahedron, both will have the same formula and the dimeric assembly will have the formulation M_2X_7.

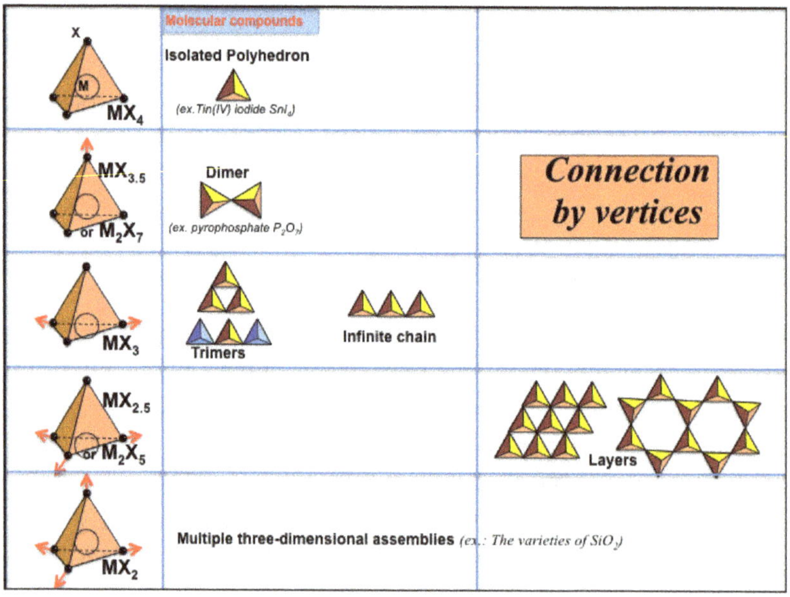

Fig. 3.32. Bonding possibilities of a tetrahedron with other polyhedra with the general formulas of the corresponding starting tetrahedron (left column). The other column represents the various configurations when the other polyhedra are also tetrahedra. From the third line ($\mathbf{MX}_3$), the associations can be either molecular or extented, depending on the fact that the tetrahedra linked to p_0 have or not the same connectivity as the starting tetrahedron. For example, if the outer tetrahedra (blue tetrahedra) only accept one bond with p_0, the association is limited to a linear trimer.

For the association of corner-sharing tetrahedra, this method leads to formulations in the range MX_4–MX_2, the topology of which is easy to represent around p_0, in the same way as the reader could do for the search of isomers of a given molecular compound.

Such an analysis is of course not limited to only one bond per X atom. Two, three or more bonds can occur with the neighboring polyhedra. The corresponding analyses do not present any difficulty. A first conclusion can therefore be extracted: whatever their nature, the coordination polyhedra, the connectivity of X and the free energy ΔG resulting from the associations are the main factors determining the topology of the frameworks and their possible polymorphs.

When the association, imagined *ex nihilo*, is only composed of one type of corner-shared polyhedra, the above method provides the *structural* formula of the framework but not the *chemical* formula. The latter is essentially a function of the oxidation number of the elements M and X. For instance, take the M_2X_5, and suppose that X are oxide ions O^{2-}. What metals can be associated with the latter, if one remembers that the solid is electrically neutral? Many solutions are possible:

— The compensation of the ten negative charges of the oxide subnetwork requires cationic couples like M^V–M^V (the upper case in Roman numbers indicates the coordination number), M^{IV}–M^{VI}, M^{III}–M^{VII} and M^{II}–M^{VIII} for electroneutrality. Unfortunately, if the polyhedra involved in the *ex nihilo* model are tetrahedra, only a few solutions chemically exist because it is rare that the corresponding cations exhibit **simultaneously** this coordination.

— If the M_2X_5 structural formula does not correspond to a neutral framework, the number of possibilities significantly increases. Imagine that the framework becomes anionic with formulae $M_2X_5^-$ or $M_2X_5^{2-}$. For reaching the electroneutrality of the resulting solid, this requires the introduction of mono- or divalent cations A (A: alkaline, alkaline-earth) in the interstices of the anionic frameworks $M_2X_5^-$ or $M_2X_5^{2-}$. The resulting chemical compositions are then $[A^+] M_2X_5^-$, $[A^{2+}][M_2X_5^{2-}]$, and $[A_2^+][M_2X_5^{2-}]$. The choice of metallic couples becomes chemically larger (Zn, Al...). To increase the number of possible combinations, one can also play on a mixture of anions on the X subnetwork (O^{2-}, F^-; O^{2-}, S^{2-}), leading to oxyfluorides and oxysulfides, respectively. For the crystal chemist, this illustrates the fact that, starting from a structural model, the chemical imagination enhances the variability of chemical compositions and, later, the range of properties.

This variability, limited when the polyhedra are tetrahedra, becomes by far much more important when octahedra (statistically the most important polyhedron for oxides, fluorides, sulfides and selenides) are considered. Instead of detailing a boring list, Figs. 3.33–3.35 present the most frequent cases encountered in the following. For this polyhedron, they correlate X connectivity, topological dimensionality and general chemical formulation. Four cases are considered: (i) octahedra sharing only vertices (Fig. 3.33); (ii) edge- and face-shared octahedra (gathered in Fig. 3.34), (iii) some cases of mixed connection (Fig. 3.35).

The part of Fig. 3.33 devoted to diverse tetramers shows that, despite the constance of the number of **M** elements, **X** is variable, depending on the conformation of the assembly of polyhedra. For a constant M, X increases when the number of shared edges decreases.

The mixed combinations are quasi-infinite. A few of them are represented in Fig. 3.35, most of the time in projection. It is clear that they develop in the third dimension, along the direction of the eye.

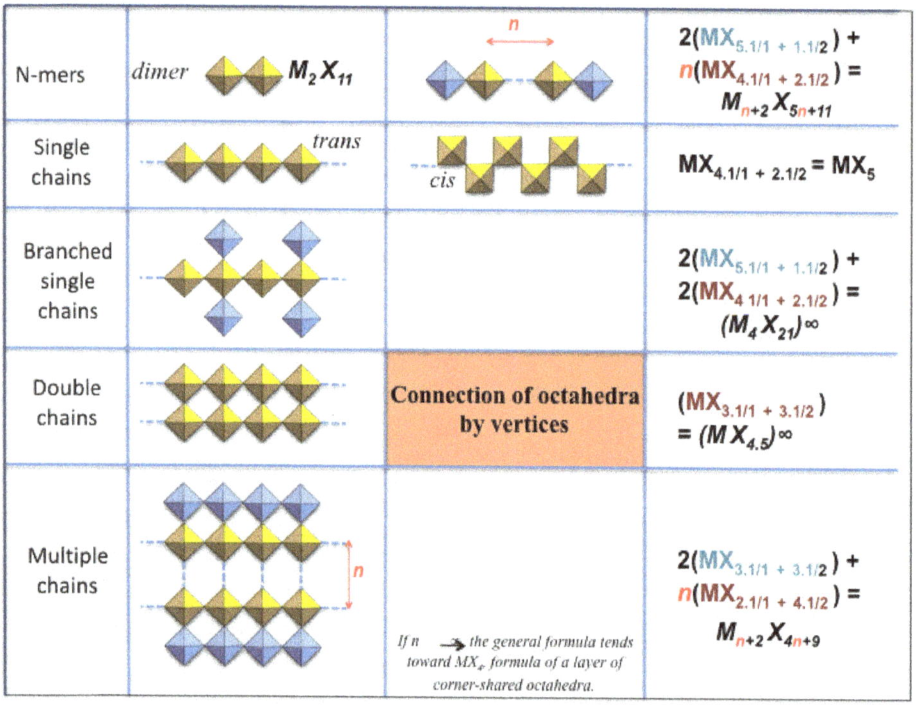

Fig. 3.33.   Formulation of some associations resulting from the connection of octahedra exclusively linked by vertices, and which present an unidimensional topology. The blue octahedra correspond to terminal octahedra; they are linked just by one vertex to the others and adopt a $MX_{5.5}$ formulation. The equilateral triangle and the squares of octahedra with formulation MX_5 were obviously omitted in the line of *n*-mers.

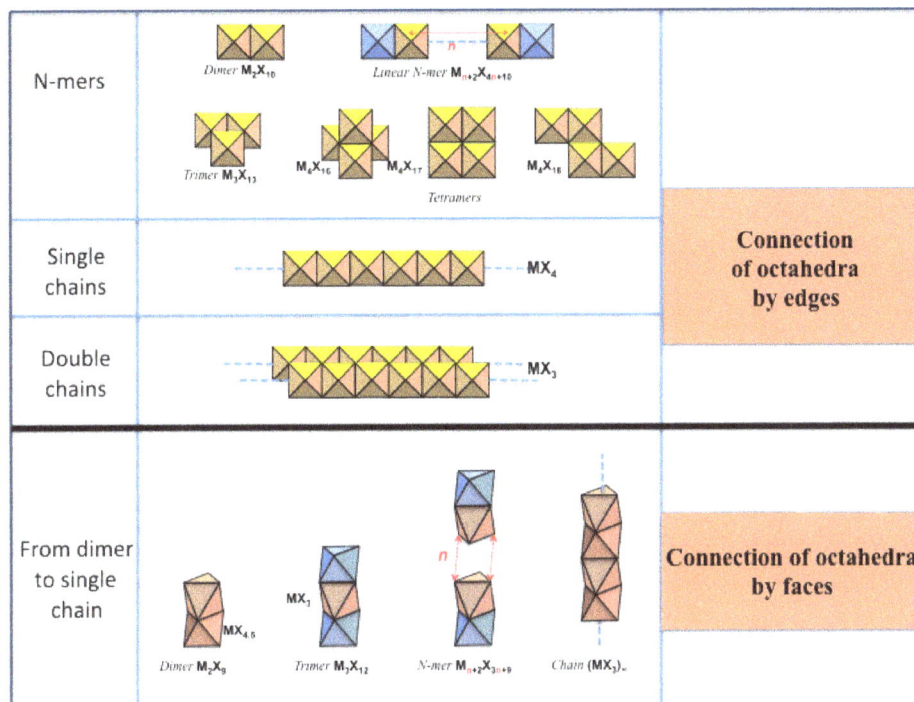

Fig. 3.34. Formulation of some associations resulting from the connection of octahedra sharing exclusively edges or faces. The resulting dimensionality is either 0 or 1. The color codes are the same as in Fig. 3.32.

Intuitively, the polyhedra described above were considered as regular. From the example of MoO_3 (Fig. 3.26), it is however not always the case. In this octahedron, a large disparity in the Mo–O distances is observed in the range $\langle 1.67–2.33 \,\text{Å} \rangle$ (2.33; 2.25; 1.95 (twice); 1.73 Å and 1.67Å). A question then arises: does the longest distances effectively belong to the polyhedron which could also be considered as a tetrahedron when distances lower than 2.00 Å are taken into consideration? The concept of **bond valence**, created by Brown and Altermatt in 1985 and further developed by M. O'Keeffe and N. Brese replies almost quantitatively to this question.

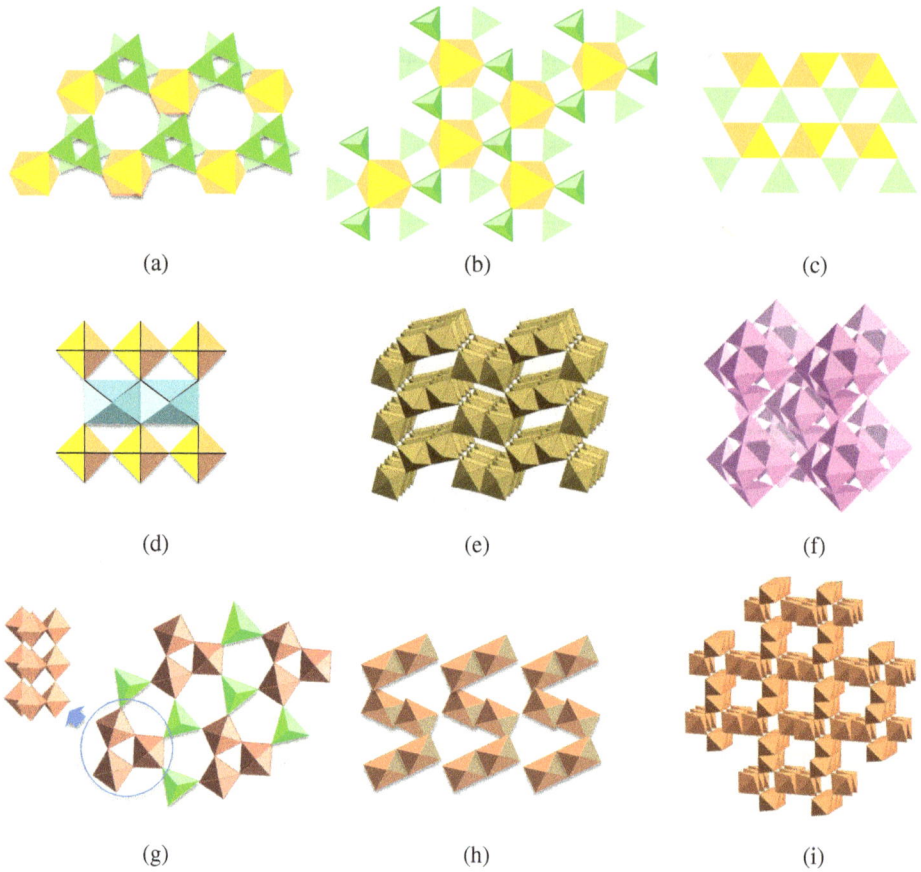

(a)   (b)   (c)

(d)   (e)   (f)

(g)   (h)   (i)

Fig. 3.35.   (a) Association by corners of tetrahedral M_3X_9 trimers with isolated octahedra; two such layers simultaneously ensure an edge-sharing connection of the trimers, and a face-sharing connection for octahedra. It is the case for the structure of the mineral benitoite $RbTaSi_3O_9$. (b) Association by corners between isolated tetrahedra and octahedra; each octahedron is linked to six tetrahedra three above the upper face of the octahedra, three below its lower plane; case of Nasicon $NaZr_2(PO_4)_3$. (c) Rutile chains (edge-sharing within them) connected to each other by corners; the corresponding layers are linked together by single tetrahedra which share three of their corners with four rutile chains (case of $CuGeO_3$). (d) Corner-sharing connection of trans-files of corner sharing octahedra with rutile chains (case of $MnCrF_5$). (e) Double chains of edge-shared octahedra linked by vertices in the two other directions ($CaTa_2O_6$). (f) Connection of square plane pyramids exclusively connected through corners; this original topology leads to super-octahedra built up of six pyramids; these super-octahedra are linked together through vertices and lead to an augmented perovskite structure (case of Nb_5F_{15}). (g) Infinite trimeric chains linked together by isolated tetrahedra ($W_3P_2O_{14}$). (h) A topology close to that of (e) with a different orientation of the double chains, due to the different sizes of the inserted cations Ca for (e); Ba for (h) ($BaNb_2O_6$). (i) Structure of the hollandite variety of MnO_2: perpendicular infinite double chains of edge-shared octahedra linked by corners; they lead to the formation of infinite tunnels with a square section.

3.5 True Coordination and the Bond Valence Method

The valence of a bond ν between two atoms i and j is defined in such a manner that the sum of all the bond valences around an atom i with valence V obeys the relation:

$$\Sigma_{ij}\, \nu_{ij} = V.$$

ν_{ij} being empirically defined as:

$$\nu_{ij} = \exp[(R_{ij} - d_0)/b].$$

In this formula, R_{ij} is the parameter of bond valence; its value was tabulated for many couples of atoms (Brese and O'Keeffe, 1992), after the examination of thousands of known structures; d_0 is the observed length of the bond and b is a constant for all atoms. Its current value is 0.37.

The most evident application of this concept will be to predict the bond lengths from these parameters. Compared to its valency, the sum of the bond valences around an atom allows the verification of the validity of a structural result. This method is much more accurate than the sum of the radii of the atoms, given before. It takes into account the distortions of polyhedra, whereas the sum of radii just provides the mean distance between atoms. These parameters are also useful for deciding if there is an effective bond between two atoms and, consequently, define the most probable coordination.

As an exercise, come back to the two structures MoO_3 and V_2O_5. The first, bidimensional, was described in detail in Sec. 3.2.6.4. Theoretically, the second (Fig. 3.36) contains vertex-shared double chains of polyhedra, connected together by sharing edges. If the polyhedron was regular, it would be an octahedron and the structure would be tridimensional. However, the determination of the structure shows a strong distortion of the polyhedra with the existence of very short $M=O$ bonds and one very long. Therefore, what is the true coordination of V in V_2O_5? Six or five?

Before any calculation, it is worthy to note that the comparison between MoO_3 and V_2O_5 proves, for each oxygen atom, different types of connection around the metal. Three of these oxygen atoms (O_1 atoms for MoO_3 and O_2 for V_2O_5) are connected to three molybdenum ions, whereas the three others (the two O_2 atoms and O_3 for MoO_3, and the two O_1 atoms and O_3 for V_2O_5) are connected only to two metals. The calculations must verify that. Indeed, the sum of the bond valences must be close to the value of the valency for each atom, cation and anion as well.

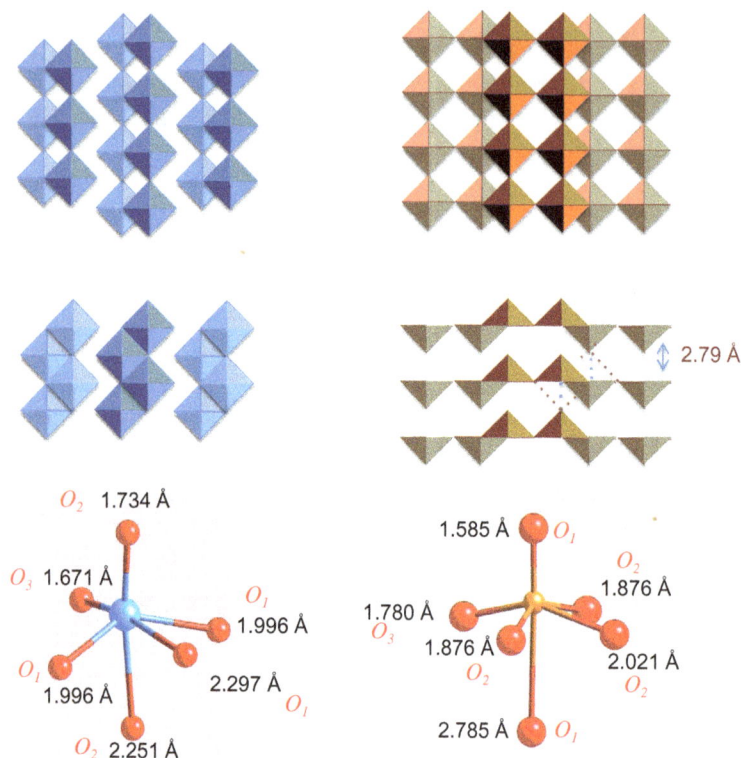

Fig. 3.36. For each compound, two perpendicular polyhedral projections prove the structural fili-ations between these two solids. On the bottom part, appears the disrtorted environment of each polyhedron, with the corresponding distances and the labels of each oxygen atom in the data tables.

From the tabulations of O'Keeffe, the values for R_{ij} are 1.803 for V^{5+} and 1.907 for Mo^{6+}. This leads to the table below.

	O_1	O_2	O_3	Σ = Valence of M
Mo	1.996 Å × 2	1.734 Å × 1	1.671 Å × 1	
	0.788 × 2	1.696	1.892	
	2.297 Å × 1	2.251 Å × 1		**5.898**
	0.339	0.395		
Σ = Valence of O	**1.915**	**2.091**	**1.892**	
V	1.585 Å × 1	1.876 Å × 2	1.680 Å	
	1.803 × 1	0.823 × 2	1.064 × 2	
	2.785 Å × 1	2.021 Å × 1		**5.240**
	0.040	0.555		
Σ = Valence of O	**1.843**	**2.091**	**2.128**	

In the lines dedicated to each metallic ion, its distances to the various oxygen sites (O_1 to O_3) surrounding it are noted in italics (in xn, n indicates the number of equal lengths for the considered atom). Below each distance, the bond valence corresponding to this M-O distance appears. The horizontal sum of these local bond valences provides the calculated valence of M (5.898 for Mo). Simultaneously, in the columns relative to each oxygen, the vertical sum of these local bond valences provides the calculated valence of O ($2 \times 0.788 + 0.339 = 1.915$ for O_1).

Three remarks emerge from this calculation.

The first relates to the importance of the weak values associated to long distances for discriminating between two types of coordination. In MoO_3, the distance Mo–O_1 (2.297 Å), even if it participates in only 6% in the valency of molybdenum, represents more than 15% of the valency of oxygen O_1, which is not negligible. This oxygen is then involved in the global coordination of Mo, which can be reasonably considered as six-fold. Application to vanadium (long V-O_1: 2.785 Å) shows that this bond valence (0.04) has almost no contribution (2%) this time to the valency of both vanadium and oxygen. The vanadium coordination is no longer octahedral by being only five-fold in the form of a square-based pyramid. Consequently, the structure of V_2O_5 is no longer three-dimensional but only two-dimensional with a Van der Waals gap, as already shown for cadmium halides.

As a second remark, this calculation is much more pertinent for the definition of the coordination than the sum of ionic radii (here 1.95 Å for MoO_3; 1.90 Å for V_2O_5) which only reflects the mean distances within the polyhedron (here $\langle 1.99$ Å$\rangle$ for MoO_3; $\langle 1.90$ Å$\rangle$ for V_2O_5).

The last remark, much more general, concerns the observed discrepancy between calculated and theoretical values for the valency of each anion. It is normal for at least two reasons: (i) the values of R_{ij} result from statistics on numerous structures, with a standard deviation which is not always negligible; (ii) the structural data are an experimental result; their accuracy depends on the quality of both the measurements and the degree of perfection of the crystal used for the determinations. They have an impact on the accuracy of the observed distances, and an error of ca. 10% is admitted compared to the theoretical valency.

Chapter 4

Some Disorder within Order! Tiltings, Vacancies and Supercells

Le désordre de ma maison reflète le désordre de ma pensée.
André Gide

Most of the structures presented in Chapter 3 were very often considered as ideal models, even if, at the end of this chapter, it was shown that distortions could exist inside the polyhedron. Beside them, anions can serve as "knee-caps" between the polyhedra. They modify the respective orientations of the latter and sometimes change the symmetry of the structural arrangement. These changes very often correspond to rotational rockings, internationally called "tiltings". This chapter presents their description, the perovskite structure being its backbone.

Despite its crystallographic simplicity (Fig. 4.1), the AMX_3 perovskite family (A: inserted cation; M: metal in octahedral environment, X being chalcogenides and halides, either alone or mixed) is chemically one of the richest structural families, due to the variety of compositions it offers in solid state chemistry.

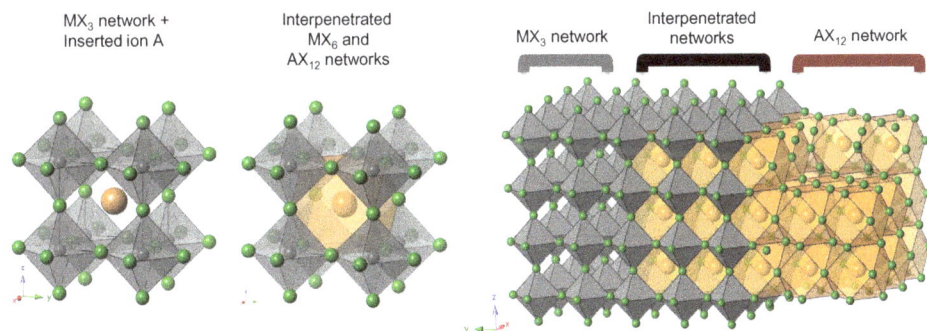

Fig. 4.1. The perovskite structure represented as a network of corner-shared octahedra (left) or as the three-dimensional interpenetration of the latter with a subnetwork of face-sharing cuboctahedra.

Indeed, both the A and M subnetworks can accept numerous valences, submitted to the global electroneutrality of their associations. The A^{n+} cation can be either mono-, di- or trivalent or also a vacancy (ex: $\square ReO_3$); at the same time, the charge of M^{p+} can oscillate between +2 and +6, always respecting the rule of electroneutrality. As the X sites are also able to accept vacancies, this obviously leads to a very large number of combinations.

The ideal perovskite structure AMX_3 can be compared to a soldier at attention, apparently — but not only! — rigid. But a soldier, as the perovskite, is not always in this position (Fig. 4.2)!

4.1 Deformations of Polyhedra

$GdFeO_3$ is a particularly significant example for this point. From the observed tiltings, the first difference observed by the eye is the change of the shape of the initial square window limited by four octahedra. It becomes a lozenge. This means that the cubic symmetry observed for $KNiF_3$ is no longer valid, and that the initial A_4 axis is now replaced by an A_2 one. As the tiltings occur in the three directions of the space, an orthorhombic symmetry is expected. It is experimentally the case.

In terms of specific vocabulary, note some new words: the ideal arrangement is called *aristotype* and the distorted one: *hettotype*. In terms of crystal chemistry, the extent of the distortion of regular octahedra can be intuitively related to the size of A cations. However, beyond tiltings, the changes of symmetry can occur even when no tilting is observed. In this case, the non-regularity of the single

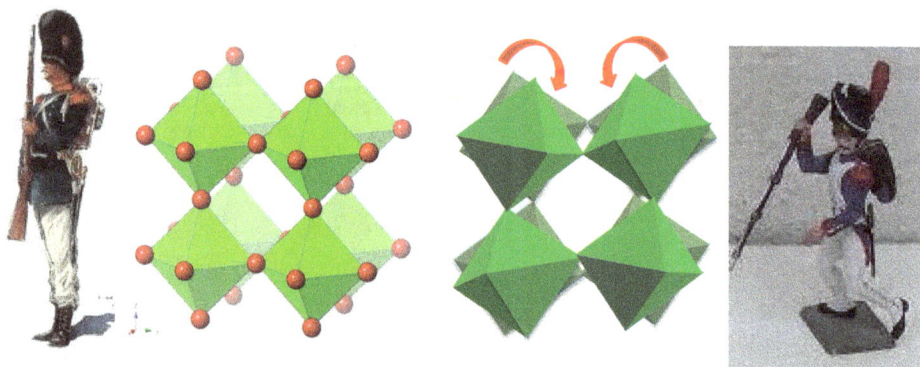

Fig. 4.2. Assuming that the octahedra are perfectly regular in both cases, perspective views of the ideal perovskite $KNiF_3$ and of the distorted $GdFeO_3$ (right) in which the octahedra present rockings (tilts) in several directions of the space. Note that the volume of the true cell of $GdFeO_3$ is eight times that of the subcell represented here.

octahedra must be taken into account. $BaTiO_3$ illustrates this situation. It is not cubic, but tetragonal.

$BaTiO_3$		space group: P $4mm$ (n° 99)	
Tetragonal	$a = 3.998$ Å	$c = 4.018$ Å	$Z = 1$
Ba 1a	0 0 z	with $z = 0$	
Ti 1b	½ ½ z	with $z = 0.482$	
O1 1b	½ ½ z	with $z = 0515$	
O2 2c	½ 0 z	with $z = 0.016$	

After the classical study described before, it is important to carefully look at a single octahedron of this structure (Fig. 4.3).

Even if the octahedron seems regular when looking at the O–O distances $((2.828$ Å$) < d(O-O) < (2.838$ Å$))$, the position of titanium is off-centered from the perfect square formed by the oxygen atoms O_2 by a distance of $(0.515–0.482) \times c$, or $0.033c = 0.132$ Å. It could at first glance be considered as negligible… It is not when looking at the Ti–O–Ti angle between two octahedra. Whereas it is strictly 180° in the perfect perovskite, it becomes 172°4 in $BaTiO_3$ [Fig. 4.3(b)]. However, the lines defined by O_1–O_1 being always parallel, there is no tilting of the octahedra in this structure. Moreover, the displacement of Ti^{4+} occurs in the same direction in two consecutive octahedra [Fig. 4.3(c)]. This leads to a polar structure, and $BaTiO_3$ is **ferroelectric**. Its polarization (noted by arrows) can be reversed by an inversion of the electric field. Once the relation between structural characteristics and physical properties is established, it appears that crystal chemistry can become a **qualitative predictive tool** for anticipating the physical properties of a new solid.

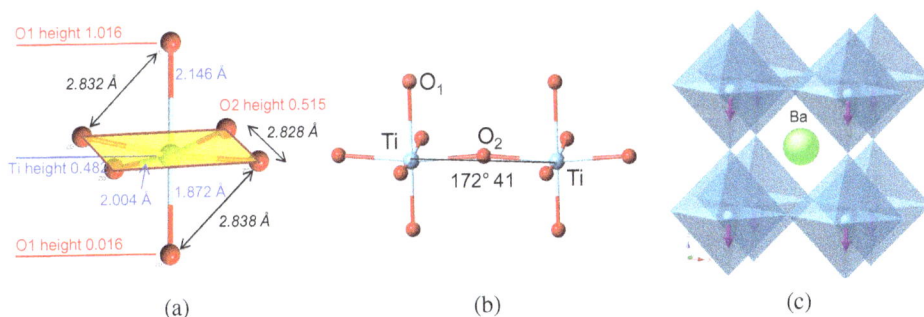

Fig. 4.3. (a) Distances in the TiO_6 octahedron; (b) non-collinearity of atoms Ti and O versus the crystallographic axes; (c) the cooperative displacement of Ti^{4+} ions is indicated by purple arrows.

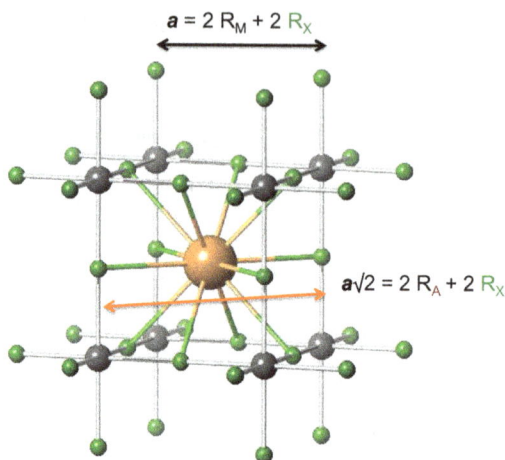

Fig. 4.4.   (a) Characteristic distances in the ideal structure AMX_3 (A: orange; M: grey; X: green).

For structures exhibiting tiltings, it was noted before that the size of cation A can play an important role which must be quantified from the examination of the ideal perovskite (Fig. 4.4). It is obvious to say that:

$$R_A + R_X = \sqrt{2} \cdot (R_M + R_X).$$

If R_M and R_X are fixed, the ratio $(R_A + R_X)/\sqrt{2} \cdot (R_M + R_X)$, called **tolerance factor of Goldschmidt** and noted t, characterizes the influence of the size of cation A.

$$t = (R_A + R_X)/\sqrt{2} \cdot (R_M + R_X).$$

It is equal to 1 for the perfect structure in which all the atoms are tangent. Depending on the respective sizes of A and M, and keeping the ideal structure, its value lies in the range 1.17 (for $CsFeF_3$)–0.807 (for $RbCaF_3$). Note that the two known structures that are the closest to ideality are $KCoF_3$ for $t >1$ (1.003) and $KZnF_3$ for $t <1$ (0.998). In each case, the resulting configuration corresponds to a minimum of the lattice energy.

4.2 Tilting of Polyhedra and Glazer's Notation

The major part of this paragraph will be dedicated to the tiltings of octahedra, considered in a first step as rigid, but not necessarily ideal (as seen with $BaTiO_3$). The tilting of other polyhedra will be developed at the end of this chapter.

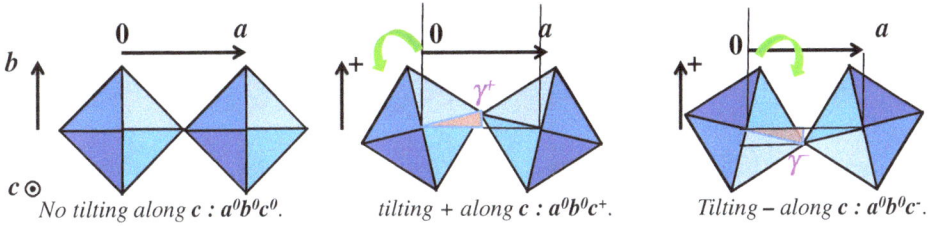

No tilting along c : $a^0b^0c^0$.   tilting + along c : $a^0b^0c^+$.   Tilting − along c : $a^0b^0c^-$.

Fig. 4.5. The two tilting modes (+ and −) of a chain from the linear chain (left).

In the real solid, tilting can occur along one, two or three directions of the reference trihedron. Its associated movements are defined by both their magnitude and the sense of rotation of the octahedron around the concerned axis. For its symbolic characterization, the notation of Glazer is used world-wide (A.M. Glazer, *Acta Crystallographica* **1972**, *B28*, p. 1384). For each axis, the movement is defined by a letter (related to the concerned rotation axis) and an exponent: +, 0 or −, indicating the sense of rotation around the axis of the polyhedron the closest to the origin (or containing it). With this convention, the ideal perovskite is noted $a^0a^0a^0$ in terms of tilting. For an illustration, take an infinite chain of octahedra along a, but tilted by a rotation around the c axis (Fig. 4.5). There are two possibilities for the sense of rotation of the octahedron in the chain.

Note that the anion serves as the knee-cap for keeping the three-dimensional (3D) topology — if one octahedron rotates in one direction, the following one will move in the other direction. Tilting is therefore a cooperative movement within a constant topology.

If this principle is applied to the complete structure instead of the chain (Fig. 4.6), a new evidence appears, related to the concept of periodicity. This property is immediately modified, depending on the sign + or −. Compared to the initial cell parameters of the ideal structure, some of those of the tilted phase are doubled due to the tilt. Consider the chains of octahedra perpendicular to the plane of the sheet in Fig. 4.6. They develop along the c axis of the cell. If the tilting is $a^0b^0c^+$, all the octahedra of this chain rotate in the same sense and project exactly one over the other. This case is called *in-phase rotation*. If the tilting is $a^0b^0c^-$, within the same chain, two consecutive octahedra rotate in opposite directions. The rotation is named *antiphase rotation*. As a conse-quence, not only the cell parameters, but also the cell volumes and the Bravais lattices are modified by reference to the ideal structure: in the case $a^0b^0c^+$, the volume is doubled; in the case $a^0b^0c^+$, the volume is four times the initial volume (see legend of Fig. 4.6).

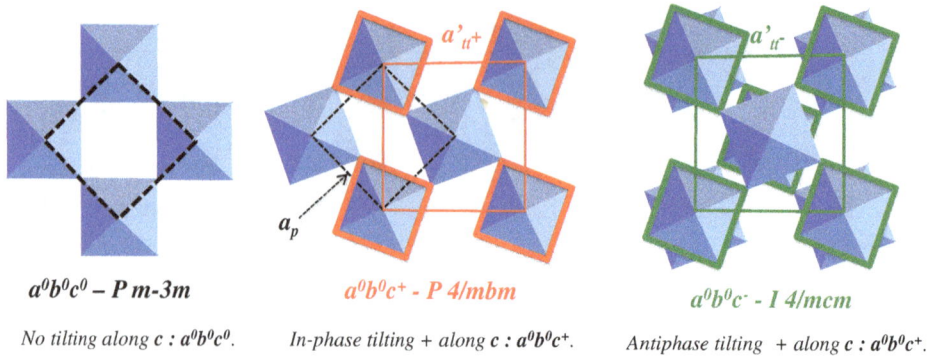

$a^0b^0c^0 - P\,m\text{-}3m$ $a^0b^0c^+ - P\,4/mbm$ $a^0b^0c^- - I\,4/mcm$

No tilting along c : $a^0b^0c^0$.   *In-phase tilting + along c : $a^0b^0c^+$.*   *Antiphase tilting + along c : $a^0b^0c^+$.*

Fig. 4.6.   (Left) The ideal perovskite: cubic cell parameter a_p, (dotted line); (middle) tilting $a^0b^0c^{+}$: all the octahedra of a chain (surrounded in red) follow an in-phase rotation; the new periodicity in the plane is characterized by the cell parameter a'_{tt^+} [for tilting], with $a'_{tt^+} = a_p\sqrt{2}$, c_p being unchanged. The cell becomes tetragonal type P, and the volume is doubled; (right): tilting $a^0b^0c^-$; $c-$ indicates that, within one chain (in green), if one octahedron rotates by an angle γ, the following one will rotate by an angle $-\gamma$ (antiphase rotation). The initial parameter c_p is doubled, and becomes $a'_{tt^-} = 2c_p$. From one plane to the other, the same orientation of octahedra is found only after a translation $\frac{1}{2}\,a'_{tt^-} + \frac{1}{2}\,b'_{tt^-} + \frac{1}{2}\,c'_{tt^-}$ which characterizes an I lattice. The cell becomes one more tetragonal (whatever the tilt, it keeps the A_4 axis) but its volume is four times the initial one.

Even if it is weak, the tilting has drastic crystallographic and physical consequences. For example, depending on the value of the M–X–M angle, the strength of the magnetic interactions between metallic ions will be determined: the larger the tilt, the weaker the magnetic interactions.

Glazer (Glazer, 1972) identified 23 possible tilts in the perovskite structure, function of the diverse rotations around the axes of the cell, with their symbol and the corresponding space group (Table 4.1).

The most common modifications of crystalline symmetry are (in decreasing order) orthorhombic (10), tetragonal (5), monoclinic (4), cubic (2). The others are isolated cases.

The various corresponding cell volumes, function of the type of tilt, are integer multiples of the ideal perovskite one. Figure 4.7 shows some of them. Depending on the type of tilting, the volume of the resulting cell can reach eight times that of the ideal one (for example, $a^0b^+c^-$, in green in the figure).

It is worthy to note that, in a given structure, the three rotations are not correlated. One can observe three different M–X–M tilting angles in the three directions of the reference trihedron.

A question then arises: *for what value of the M–X–M angle (θ_{max}) is the tilting maximum?* It is when the compacity of the X subnetwork will be maximum. This occurs when one X atom of an octahedron will project on the face of its neighboring octahedron. In this case, the resulting four X atoms, now tangent, will form a

Table 4.1.   The 23 types of tilting, with their Glazer symbols and their space group. In the 3rd column, the letters in brackets mention the adopted crystalline system: [C] cubic; [T] tetragonal; [O] orthorhombic; [R] rhombohedral; [M] monoclinic; [TC] triclinic.

Type of tilt	Symbol	Space group
Tilt 3D		
1	$a^+b^+c^+$	I mmm [O]
2	$a^+b^+b^+$	I mmm [O]
3	$a^+a^+a^+$	I m -3 [C]
4	$a^+b^+c^-$	P mmn [O]
5	$a^+a^+c^-$	P 4_2/nmc [T]
6	$a^+b^+b^-$	P mmn [O]
7	$a^+a^+a^-$	P 4_2/nmc [T]
8	$a^+b^-c^-$	P 2_1/m [M]
9	$a^+a^-c^-$	P 2_1/m [M]
10	$a^+b^-b^-$	P nma [O]
11	$a^+a^-a^-$	P nma [O]
12	$a^-b^-c^-$	F -1 (P -1)* [TC]
13	$a^-b^-c^-$	I 2/a (C 2/c)* [M]
14	$a^-a^-a^-$	R -3c [R]
Tilt 2D		
15	$a^0b^+c^+$	I mmm [O]
16	$a^0b^+b^+$	I 4/mmm [T]
17	$a^0b^+c^-$	C mcm [O]
18	$a^0b^+b^-$	C mcm [O]
19	$a^0b^-c^-$	I 2/m (C 2/m)* [M]
20	$a^0b^-b^-$	Imma [O]
Tilt 1D		
21	$a^0a^0c^+$	P 4/mbm [T]
22	$a^0a^0c^-$	I 4/mcm [T]
Tilt 0D		
23	$a^0a^0a^0$	P m -3m [C]

• The tilts marked with a star correspond to non-standard space groups.

The standard group of the International Tables is indicated in parentheses.

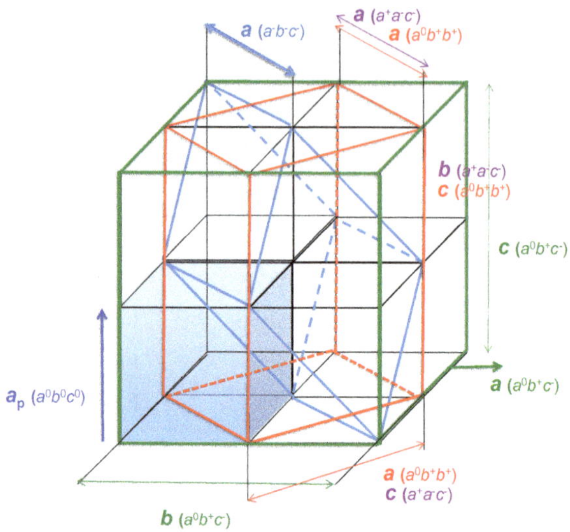

Fig. 4.7.   Some orientations of the tilted cells in relation with the ideal perovskite one (tilting $a^0b^0c^0$).

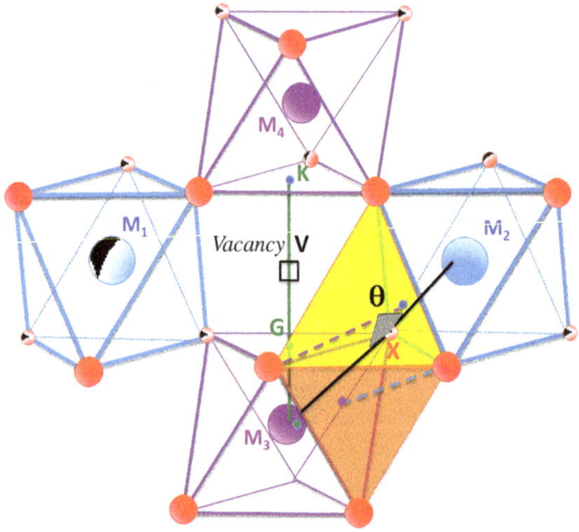

Fig. 4.8.   Position of the octahedra when the tilt is maximum. For the two consecutive octahedra, the two dotted lines show how the X atom is positioned versus the three X atoms of a face of the other octahedron in order to create an empty tetrahedron (in orange). This movement leads to the formation of planar triangular subnetwork of X atoms and to the creation of a perfect octahedral vacancy centered on V. The initial cuboctahedron becomes highly distorted.

tetrahedron (Fig. 4.8) between the two octahedra. The initial perovskite parameters are then doubled in the three directions of the space.

As the M_3–V–M_2 angle is 90°, the M_3VM_2 triangle is rectangle, it becomes easy to calculate θ_{max} from the calculation of the M_2–M_3 distance if a is the length of the edge of the (supposed regular) octahedron. By applying the formula relative to the octahedron mentioned in Appendix 1, we get

$$M_3V = 2\left(a\sqrt{6}\right)/6, \quad M_2V = a \quad \text{and} \quad M_2X = M_3X = \left(a\sqrt{2}\right)/2.$$

This leads to $(M_2M_3)^2 = 5a^2/3$.

The injection of this value in the general formulae of the metric relations in the triangle

$$a^2 = b^2 + c^2 - 2\,bc \cdot \cos A$$

gives here

$$\cos \theta = -2/3$$

and

$$\boldsymbol{\theta = 131°81}$$

instead of 180° in the ideal perovskite. This large difference shows how much the perovskite structure is adaptable according to the choice and the size of its cations. Figure 4.8 also shows that the tilting model is in this case $\mathbf{a^-b^-c^-}$ through anti-phase rotations. Two different distortions correspond to this configuration: one is triclinic (type 12 of Table 4.1), the other is monoclinic (type 13). Due to the existence of an angle of 90° in Fig. 4.8, the monoclinic situation is represented. Under these conditions, it is even possible to calculate by the same general formula the monoclinic β angle of the subcell defined by $M_1M_2M_3M_4$. A value of 101°53 is found for β. The space group is C 2/c, as proposed by Glazer (Fig. 4.9).

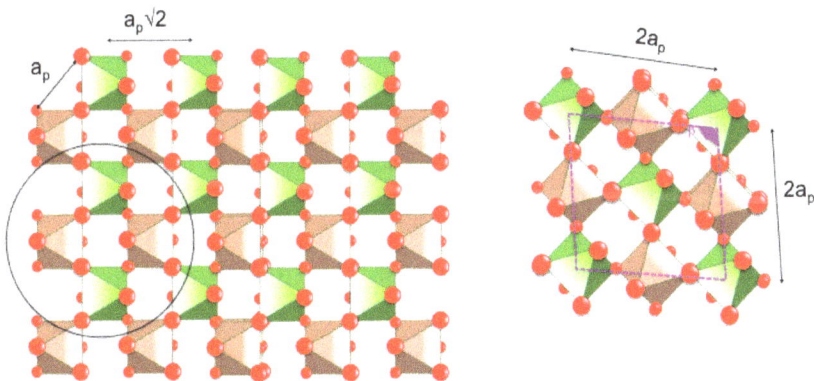

Fig. 4.9. Two extented projections of the cell in Fig. 4.8, tilted along z (appearing with a 90° rotation inside the circle on the left and, on the right a 45° rotation with the β angle). The two colors differentiate the rockings. As in Fig. 4.8, the size of anions X takes into account the proximity to the observer: the larger the size, the closer the observer.

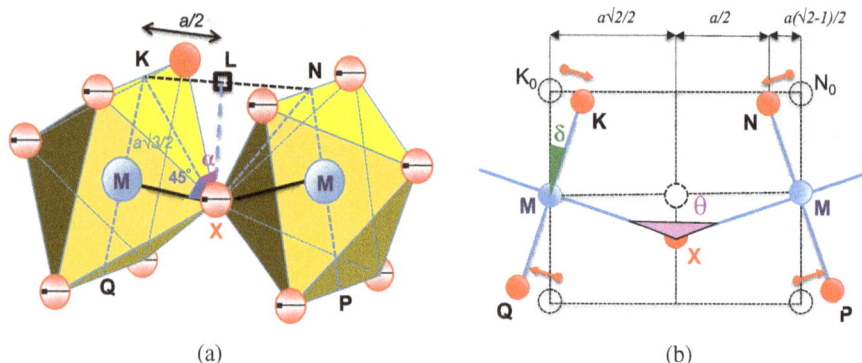

Fig. 4.10.   (a) Perspective view of the rocking of two octahedra tangenting their edge; (b) the same in projection for estimating δ, the angle of rocking during the tilt. The dotted lines indicate the situation before the tilt.

This represents an extreme case. There is another one, less important, when the anions X on the edges of two adjacent octahedra become tangent (Fig. 4.10) by simultaneous rotation around the anion X shared between the two octahedra. Following the information observed in Fig. 4.10, the θ value of the M–X–M tilting angle is deduced from the value of the α angle, since $\theta = 2(45° + \alpha)$. An easy calculation leads to $\alpha = 35°26$, and $\theta = 160°52$. Finally, as the distance between the X atoms of the two different edges (tangent after the tilt) decreases from $a\sqrt{2}$ to a (a being always the length of the height of the octahedron), the value of the rocking angle δ (11°95) is easily accessible.

The two examples above give an idea of the wide range of distortions $(180° > \theta > 132°)$ which can be encountered when structures are built from corner-sharing octahedra, depending on the characteristics of the cations A and M (or A and A', M and M' when a mixture of metals fills the octahedra and cuboctahedra).

Even if the case will not be treated in this book, it is worthy to note that edge-shared octahedra (meaning M–X–M angles close to 90°) can exhibit tiltings, but the authorized range is much more restricted (only a few degrees). On the contrary, no tilting is observed when octahedra share faces. (M–X–M angle strictly equal to 70°50.) A similar calculation can be performed with tetrahedra. When they share vertices, the magnitude of the tilting lies in the range $(102° \leq \theta \leq 180°)$; sharing edges correspond to $66° \leq \theta \leq 70°50$; for shared faces, no tilting (M–X–M angle strictly equal to 39°).

Along with the 23 possibilities of tilting of Glazer, the extreme case $\mathbf{a^-a^-a^-}$ deserves attention. It concerns only anti-phase rotations. The symmetry is

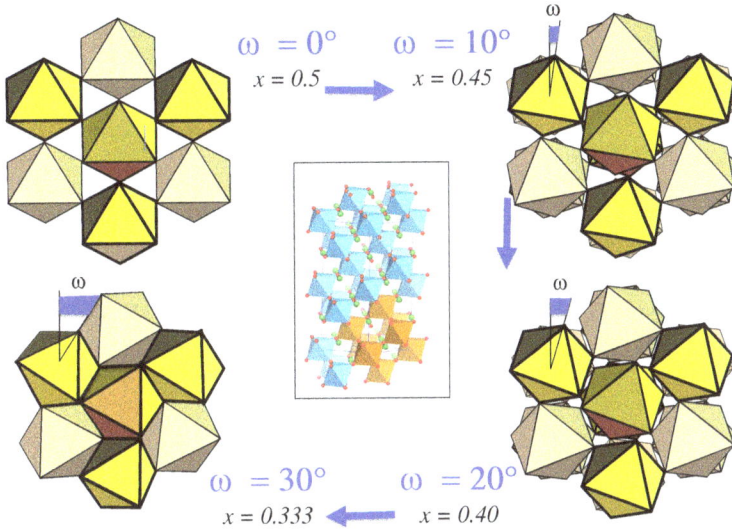

Fig. 4.11. Changes of the orientation of the octahedra by rotation around the *z* axis of the hexagonal super-cell, depending on the value of the *x* coordinate of the anions; this rotation is characterized by the angle *ω*. In the insert, the octahedra in orange reminds the reader of the perovskite structure viewed along this direction.

rhombohedral R–3c (Table 4.1) (a rhombohedron is obtained from a cube by elongation along one diagonal of the cube). Referring to International Tables shows that, treated in the hexagonal multiple cells of R–3c, A ions are in (0, 0, ¼) positions, B cations in (0, 0, 0) and the anions in (*x*, 0, ¼). The value of *x* determines the magnitude of the tilt (Fig. 4.11).

In this particular case, the angle of tilting, labelled *ω*, is not defined as usual, around the crystallographic axes, but around the axis perpendicular to the face of an octhedron. Its value is a function of the reduced coordinate *x* of the anion. It is easy to show that:

$$\tan \omega = \sqrt{3} - (x\sqrt{12}) \quad \text{and also} \quad x = 0.5 - [\tan \omega / 2\sqrt{3}]$$

ω varies from 0° for *x* = 0.5 to 30° when *x* = 1/3.

Except for this special case, the tiltings are defined from the crystallographic axes. For illustration, Fig. 4.12 presents two examples of tilting: $a^+a^+a^+$ (orthorhombic I mmm) and $a^+a^+c^-$ (tetragonal P $4_2/n$). They are encountered in metallic hydroxides. In the first, a hydrated ferric hydroxide, three in-phase tiltings are observed. If a_p is the length of the initial cell parameter of the ideal perovskite, the new volume is defined by $2a_p \times 2a_p \times a_p$, representing a four-times larger cell

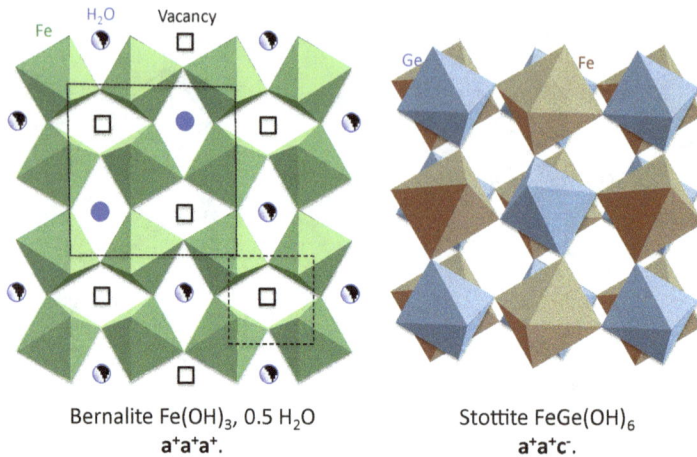

Fig. 4.12. Two different examples of tilting modes induced by cationic ordering (Fe^{3+}: green; Fe^{2+}: pale brown; Ge^{4+}: blue; (see text)). The original perovskite cell appears at the bottom of the left figure as dotted lines.

volume. The latter is doubled for hydroxide $FeGe(OH)_6$ due to an anti-phase rotation along the c axis.

Besides the illustration of tilting by these two examples, the advantage is the introduction of two other particularities: cationic order and vacancies. In the mineral bernalite, a hydrated $Fe(OH)_3$, $0.5H_2O$ (Birch, 1993), all the metal sites B are occupied by the same cation iron(III). The water molecules occupy one-half of the centers of the cuboctahedra distorted by the tilting, in an ordered manner, the other half corresponding to vacancies (squares in Fig. 4.12). The formula of this hydroxide can therefore be written $(H_2O)_{0.5}\square Fe(OH)_3$. In the second structure (the mineral stottite $FeGe(OH)_6$ (Ross *et al.*, 1988), all the cuboctahedral sites are empty but this time the octahedral B sites are occupied by two kinds of cations (large Fe^{2+} (r: 0.92 Å) and small Ge^{4+} (r: 0.67 Å)) arranged in an ordered way. The different sizes of their ionic radii leads to the cationic ordering being imposed on the tilting. The corresponding cell volume is this time eight times that of the ideal perovskite. It is the first consequence. The second is the noticeable change of the position of the OH^- anions induced by the different size of the cations. The possibilities of cationic ordering are numerous and lead to new distortions of the perovskite structure.

4.3 Cationic Ordering and Supercells

Except for Stottite, the above examples considered that octahedra were regular and filled by only one kind of M cations. During the study of dense packings, it was shown (and verified experimentally) that the interstices within the octahedra were sufficiently large ($r/R = 0.414$, *cf.* Chapter 1, p.25) for hosting a variety of cations. For a given charge for the anion X, a variety of A–M cationic couples could adopt the perovskite structure. For instance, when X is oxygen, one can associate:

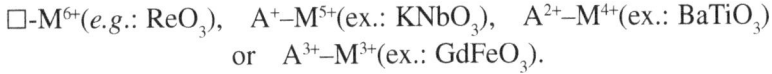

$$\square\text{-M}^{6+}(e.g.: \text{ReO}_3), \quad \text{A}^+\text{-M}^{5+}(\text{ex.: KNbO}_3), \quad \text{A}^{2+}\text{-M}^{4+}(\text{ex.: BaTiO}_3)$$
$$\text{or} \quad \text{A}^{3+}\text{-M}^{3+}(\text{ex.: GdFeO}_3).$$

Always keeping the electroneutrality rule for the solid, one can also increase the richness of the perovskite family by introducing simultaneously several M^{p+} and/ or M^{n+} ions of different sizes. Nature did that beautifully far before the chemists, as exemplified by looking at the composition of the natural minerals!... For example, the chemical composition of the loparite mineral, which exhibits a ABX_3 perovskite topology, is: $(Ce,Na,Ca)(Ti,Nb)O_3$: three types of cations with mono- di- and trivalency on the A site, two for B site, the general formula always being ABX_3. The structural adaptability and the chemical variability on each site of the perovskite inspired a lot of chemists. They created new families by synthesizing solids with the $[(A,A')(B,B')X_3]$ formula, for different values of the ratios between the cations on each site. As chemists have a Cartesian philosophy, their first approach was to study special compositions corresponding to simple values of the ratios: $\frac{1}{1}$, $\frac{1}{2}$, $\frac{1}{3}$...

From their studies, two different situations emerge: (i) for example, a random distribution of A and A' and/or B and B' cations on the site; (ii) an ordered repartition.

- In the first case, the statistical disorder does not fundamentally modify the structure, except a small variation of the cubic cell parameter, due to the different sizes of the two cations. Topologically, this merging effect leads to the ideal structure.
- The second case (long range order) is much more interesting. The different sizes of the mixed cations on each site induce modifications of the tiltings and, correlatively, affect the dimensions of the cell, with a large possibility of situations, depending on the nature and the ordering of the cations. Three simple examples of ordering in the **non-tilted** $[(A,A')(B,B')X_3]$ family are illustrated in Fig. 4.13. In the three cases, the specific type of ordering obliges to double each cell parameter, and the cell to be eight times that of the ideal perovskite. Three different types of ordering are observed: occupation in $(NH_4)_3FeF_6$ of the A and B sites by the same cation ion (NH_4) (Minder, 1937); order M–$\square$ on the B site (Pt–$\square$ in K_2PtCl_6 (Wyckoff, 1968)), strict order on the two sublattices (Na and Al in K_2NaAlF_6 (Morss, 1974)).

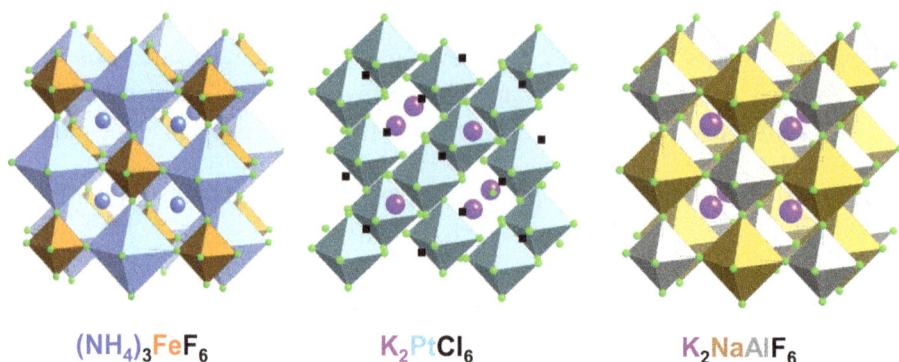

$(NH_4)_3FeF_6$ K_2PtCl_6 K_2NaAlF_6

Fig. 4.13. In $(NH_4)_3FeF_6$, the ammonium ion occupies simultaneously all the cuboctahedral A sites (dark blue circles) and one-half of the octahedral B sites (pale blue), the other half being occupied in an ordered manner by iron(III) (orange); this solid can be written $[(NH_4)_2 (NH_4,Fe)]F_6$, for identifying A and B sites; note the influence of the (NH_4/Fe) order within the B sites on the change in the positions of the fluoride ions of the anionic subnetwork. In K_2PtCl_6, the cuboctahedral A sites are filled only by K^+ ions; the tetravalency of platinum implies that the B site, due to the electroneutrality of the framework, is only half-occupied in an ordered way by Pt^{4+}, the other half corresponding to ordered vacancies (black squares) [formula $(K_2)(\square,Pt)]Cl_6$]. With the same notation, the last solid must be written $[(K_2)(NaAl)]F_6$. It corresponds to the natural mineral elpasolithe. Note that the colors of the letters in the formula below the figures recall those appearing in the figures.

The values of the ionic radii for various coordinations (in Roman exponents) of cations A and B are well known: $^{XII}NH_4^+$ (1.37 Å), $^{XII}K^+$ (1.78 Å), $^{VI}Na^+$ (1.16 Å), $^{VI}Al^{3+}$ (0.675 Å), $^{VI}Fe^{3+}$ (0.785 Å), $^{VI}Pt^{4+}$ (0.765 Å), $^{II}F^-$ (1.145 Å) and $^{II}Cl^-$ (1.67 Å). For calculating the Goldschmidt tolerance factors in these compounds, its sometimes necessary to merge the values (in brackets $\langle \rangle$) of the ionic radii when two species occupy the same sites. This gives the B/A ratios: $Fe^{3+}/NH_4^+ = 0.573$, $Pt^{4+}/K^+ = 0.430$, $\langle Na^+,Al^{3+}\rangle/K^+ = 0.515$; the Goldschmidt tolerance factors t are 0.921, 1.002 and 1.003, respectively. In the two last cases, t is close to the ideal value of 1, not for $(NH_4)_3FeF_6$, in which can be anticipated a lower stability and more constraints. Note that if potassium could replace ammonium in an ordered manner in this compound, t would become 1.070 in the case of an ordered repartition and 0.995 when a random 1:1 repartition is present. The chosen alternative depends on the calculated lattice energy in each case. The minimum energy is always favored.

In a general way, the cationic ordering is frequent and not restricted to perovskites. As for tiltings, this often leads to various super-cells, depending on the type or order and on the ratios between the species within a site. In our previous examples on tilted structures, only the 1:1 order was considered. Many others exist. Look at the rutile structure type MX_2 (Fig. 4.14). It remains unchanged when two different cations randomly occupy the octahedra (e.g. $(Fe,Nb)O_2$).

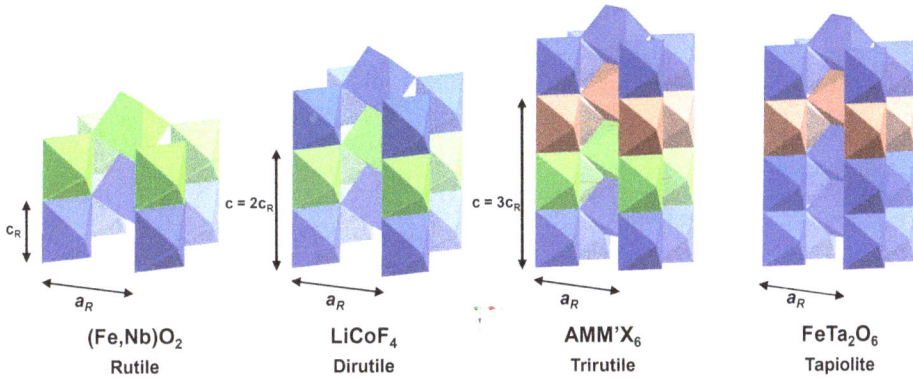

Fig. 4.14. The surstructures of the rutile structure. In dirutile, (1:1 cationic order, doubling of the *c* parameter), the *a* parameter remains identical. In trirutile, the order can be 1:1:1 or 1:2 (tapiolite). See text.

A 1:1 cationic order in the chains leads to dirutile $MM'X_4$, with a doubling of the *c* parameter (e.g. $LiCoF_4$ Lacorre *et al.*, 1989). A three-time increase of the *c* parameter is observed in two different cases of ordering in the trirutile family $MM'M''X_6$: (i) order 1:1:1 between the three cationic species if $M \neq M' \neq M''$ or (ii) order 2:1 with the same tripling if $M \neq M' = M''$ (case of mineral tapiolite $FeTa_2O_6$ (Wyckoff, 1968)).

4.4 Vacancies, Extented Defects and Non-Stoichiometry

Up to now, the vision of the perfect solid developed here considers the solid as built up of atoms occupying ideal positions in the framework. Most of the time, it is true, in the absence of any energetic external stimulus (temperature, pressure, irradiation, oxydo-reduction…). The action of the stimulus can create the conditions for a modification of the structure, corresponding to a lower minimum of lattice energy. This can give rise to atomic displacements either around their equilibrium position or from one position to another one. The most common stimulus? It is temperature which can create several types of defects in the framework. A defect is therefore an alteration of either the electronic structure or of the crystalline organization of the perfect solid. Defects are of two kinds:

(i) **The intrinsic defects** *or physical defects.* They have no influence on the chemical composition. They are also labelled *stoichiometric.*
(ii) **The extrinsic defects** *or chemical defects.* In general, they lead to modifications of the chemical composition, even if the topology is preserved.

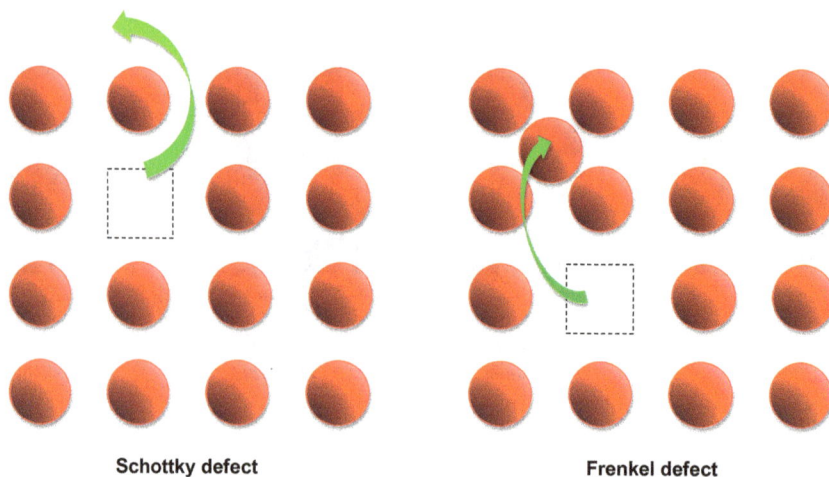

Fig. 4.15.   Reminder of the definitions of Schottky and Frenkel defects.

When the defects appear locally on random sites of the crystal, they are called **point defects**. When they are ordered, they generate **extended defects** that are reponsible for a non-stoechiometry mechanism, a phenomenon briefly considered in Chapter 3 (Fig. 3.22) regarding the quarrel between Dalton and Berthollet.

The intrinsic defects are often described in Metallurgy courses. The two most well known are those of Schottky and Frenkel (Fig. 4.15). In the first, one atom leaves its normal position, migrates to the surface, then leaving a vacancy. The Frenkel's defect corresponds to the case where the atom moves from its initial position to reach an *intersticial site* of the framework. The two kinds of defects can coexist in the same solid, even if not isolated. Cases are known where, for example, associations of vacancies (bivacancies) are more stable than two isolated vacancies.

This paragraph will mainly focus on the concept of **extended defects**, pioneered in 1948 by Arne Magnéli (Magnéli, 1948), after a long quarrel with his community which, at that time, casted some doubts on this revolutionary concept, proven by electron diffraction experiments.

4.4.1 *The extended defects*

They are of two kinds: (i) those whose transient existence will be shortened through structural rearrangements (*crystallographic shears*) which eliminate the defects, even if the resulting solid keeps the memory of their short existence; (ii) those which subsist with time as cationic and/or anionic vacancies observed during structural studies.

These defects are especially observed in *berthollides* which often have variable chemical compositions in which the cation/anion ratios are not simple. The formulae are expressed as a function of the variable x which depends on the physical conditions imposed by chemists (historical example: $Na_x W^{5+}_x W^{6+}_{1-x} O_3$ for the first Magnéli bronze (Magnéli, 1949a, 1949b)).

The first case, associated with structural modifications for eliminating the defects, can be illustrated by the example of an infinite double chain of octahedra, reminiscent of the perovskite type (Fig. 4.16).

When reduced in special conditions (for instance by hydrogen), some of the oxygen atoms of these chains involved in the linkage of octahedra will disappear in the form of water molecules. Due to the correlative new five-fold coordination, the resulting defects induce a structural instability and an augmentation of the free energy of the system. To lower it further, the framework eliminates the defects by a modification of the organization: by a crystallographic shear, it periodically

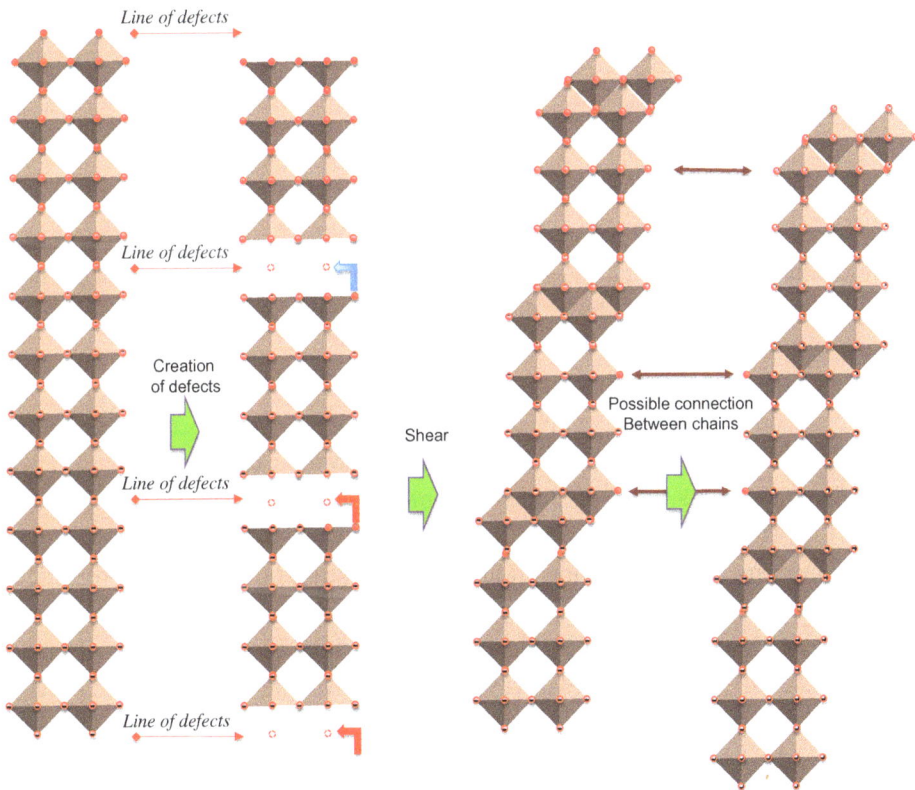

Fig. 4.16. The different steps leading to the shear in the original double chains: chemical creation of extended defects followed by shears generating periodic sharing of edges. Defects appear as empty circles and shearing movements as red arrows.

Series Mo$_n$O$_{3n-1}$ n = 8 and 9   A. Magnéli, Acta Chem. Scand. 1948, 2, 501.

Mo$_8$O$_{23}$ n = 8   **Mo$_9$O$_{26}$ n = 9**

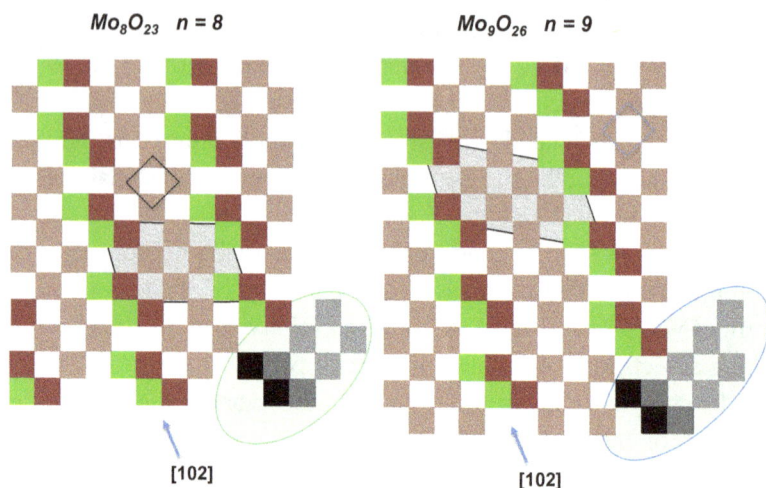

[102]   [102]

Fig. 4.17. [010] projection of the terms $n = 8$ and $n = 9$ of the Mo$_n$O$_{3n-1}$ family of the Magnéli phases. The initial perovskite cell is indicated as an empty square and the real cells as grey parallelograms. The blocks of four octahedra are aligned along the [102] rows of the initial cell. The motifs Mo$_8$ and Mo$_9$ are identified within the pale green ellipses.

brings closer together the oxygen layers around the line of defects. This regenerates the octahedral coordination of the metallic ions. The consequence is the periodic formation of blocks of four edge-shared octahedra, whereas they were linked at the beginning by vertices.

The chain, primitively linear, becomes periodically sheared due to the disappearance of one oxygen per octahedron. For an assembly of n octahedra, the formula becomes in the present case M$_n$O$_{3n-1}$. It is also the case for the two series of molybdenum oxides and of tungsteno-molybdates Mo$_{1-x}$W$_x$O$_{3n-1}$ (x small) (Fig. 4.17) (Magnéli, 1948).

The movements of the chains induce the formation of rectangular interstices. Their volume is twice that of the initial cuboctahedron. The same shear mechanism [Fig. 4.18(a)] occurs when planes are taken into account instead of chains [Figs. 4.18(c)–4.18(f)], but in two ways, depending on the localization of defects: in the square section of the octahedron [Figs. 4.18(c)–4.18(e)] or in apical position [Figs. 4.18(d)–4.18(f)].

The *shear* process is general and not limited to the formation of four octahedra blocks. In the same way, with triple chains, six octahedra blocks could be obtained, implying the elimination of two octahedra instead of one. The corresponding formula: M$_n$O$_{3n-2}$ is found in W$_{20}$O$_{58}$, which belongs to the series of tungstates starting at $n = 20$.

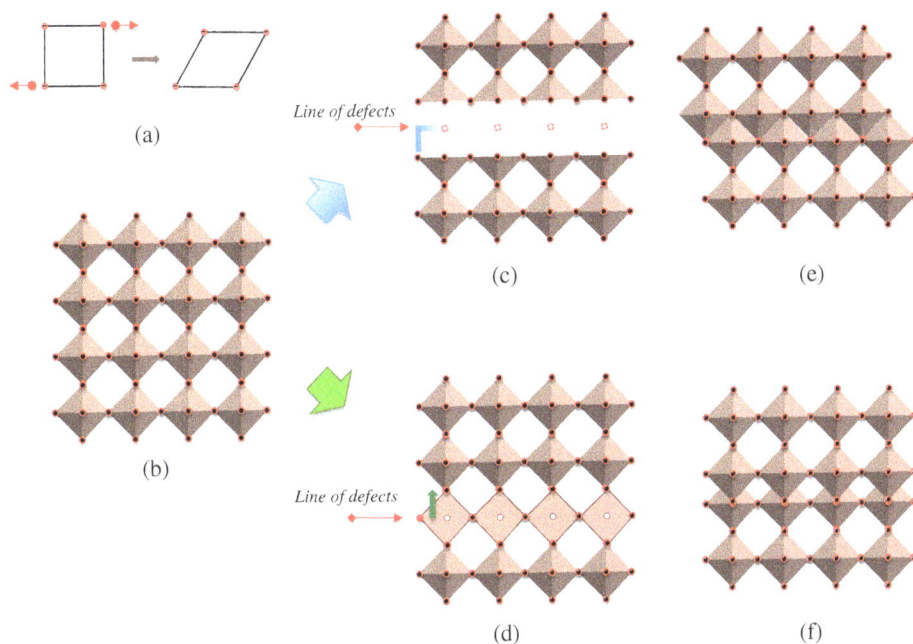

Fig. 4.18. Extension of the shear mechanism to a ReO_3-type structure. The figures on the right represent the projection along two perpendicular axes.

Moreover, these shears are not limited to one dimension. They exist in three dimensions with the formation of square or rectangular blocks of variable size. Wadsley was the first to observe this phenomenon from the high resolution electron microscopy study of heavy metals oxides, particularly niobium oxide which presents many polymorphs (Paving 1). In it, the size of each square or rectangular block (noted $[n \cdot n]$ for squares and $[m \cdot n]$ for rectangles) determines the symmetry of the resulting cell.

The ReO_3 topology is not the only one in which shear mechanisms are observed. The same phenomenon was also thoroughly studied for the rutile topology. The general formula is M_nO_{2n-1} in this case. The system, particularly rich for vanadium ($\infty \geq n \geq 3$) and titanium ($\infty \geq n \geq 4$) phases, was less studied when the metal was chromium, manganese or lead, the lone pair of the latter playing a complex role. Despite different symmetries versus n, the cell parameters of all the phases exhibit clear metric relations with the original tetragonal rutile. The shear generates not only edge-shared octahedra, but also shared faces in the blocks.

This concept, that governs (and always governs!) many parts of modern solid state chemistry, has led to a huge number of new phases. However, it is worthy to note that, for a given mechanism, it is the chemical nature of the metallic ion and its valency which will influence the final arrangements... Why Mo_nO_{3n-1} for

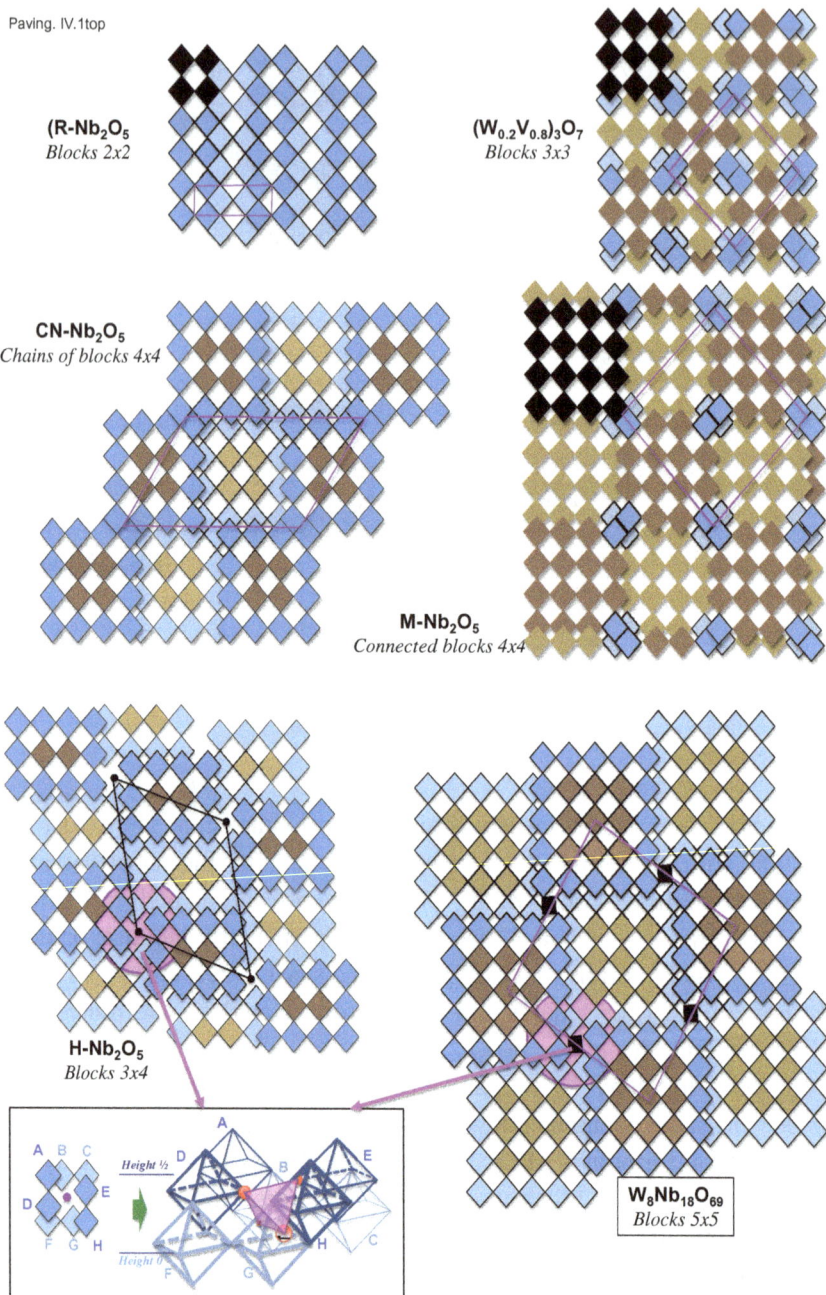

Paving. IV.1top

(R-Nb$_2$O$_5$)
Blocks 2x2

(W$_{0.2}$V$_{0.8}$)$_3$O$_7$
Blocks 3x3

CN-Nb$_2$O$_5$
Chains of blocks 4x4

M-Nb$_2$O$_5$
Connected blocks 4x4

H-Nb$_2$O$_5$
Blocks 3x4

A B C
Height ½
D
E
F G H
Height 0

W$_8$Nb$_{18}$O$_{69}$
Blocks 5x5

Paving 1. Representing in projection of the shears occurring in heavy metal-based oxides, single or mixed. In particular, the four allotropic varieties of niobium pentoxide Nb$_2$O$_5$ are presented. They are built from [n · n] blocks (2 ≤ n ≤ 4), except for the H form in which the blocks are [3 × 4]. The blue color is for layers at height ½ and the orange one for those at height 0. In the latter, orange is replaced by light purple for a better understanding of shear walls. In the two last cases, blocks at the same height are disconnected and create a tetrahedral site, presented in perspective in the inset of the bottom left part. Cells are indicated.

molybdenum? Why W_nO_{3n-2} for tungsten? Why such a passage from one composition to the other when varying the Mo/W ratios? Despite the lack of current explanation, it seems clear that the electronic structure of the metal plays a role. This parameter, joined with the diversity of the observed structural arrangements, conditions a lot of physical properties (particularly electrical) and the numerous advances that this discipline obtained during the last 30 years.

4.4.2 *Ordered extended defects; examples*

The concept of **ordered extended defects** was briefly recalled with YBaCuO in Chapter 3 and with K_2PtCl_6 in this chapter. In this case, the defects stay in the structure. There is not, this time, the need of a structural change for eliminating the defects, as before. In other words, the defects do not modify enough the free energy of the system leading to local changes in the arrangements. This situation is frequent and the examples are innumerable. Just two examples will be presented for the illustration of the structural consequences of the existence of ordered extended defects (Fig. 4.19):

Brownmillerite (Ca_2AlFeO_5) is a natural mineral [Fig. 4.19(left)] (Colville and Geller, 1971; Bertaut *et al.*, 1959). At first sight, it can be described as the periodic stacking of perovskite-like monolayers and double files of tetrahedra. In this structure type, both metal sites accept several types of transition metals (Fe, Co...), ordered or not. Whereas a Al/Fe ordering was expected in Ca_2AlFeO_5 with Al in tetrahedra and iron in octahedra, a partial disorder is observed. On the contrary, in

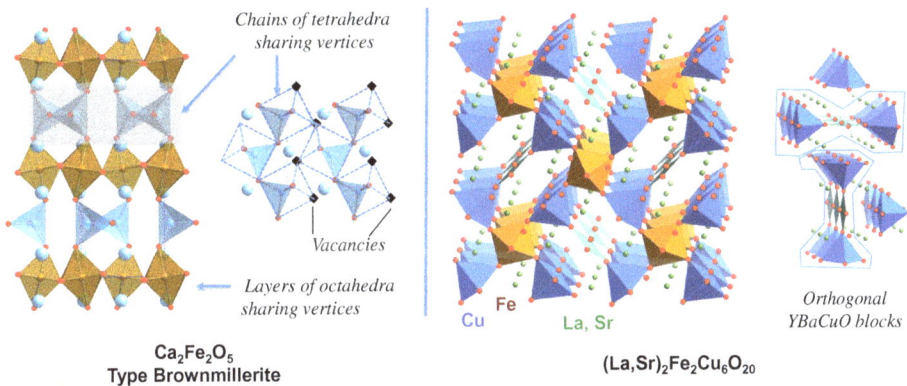

Chains of tetrahedra sharing vertices

Vacancies

Layers of octahedra sharing vertices

$Ca_2Fe_2O_5$
Type Brownmillerite

Fe
Cu   La, Sr

$(La,Sr)_2Fe_2Cu_6O_{20}$

Orthogonal YBaCuO blocks

Fig. 4.19. (Left) [001 projection] of brownmillerite, with the detail of the chains of tetrahedra which are tetrahedra which result from O vacancies in the primitive octahedra (dotted lines) with vacancies (Fe: orange, Cu: blue, vacancies: black spheres); (right) the pseudo-lacunar copper–iron-based compound (Genouel *et al.*, 1995), built from Cu^V-Cu^{IV}-Cu^V trimers, as for the YBaCuO structure. They are linked together through iron(III) chains of corner-shared octahedra.

the isotypic mineral (srebrodolskite $Ca_2Fe_2O_5$ (Bertaut *et al.*, 1959)), the same metal (Fe), only with a 3+ valency, occupies both sites.

A better look at the tetrahedral layers leads to consider that their empty spaces correspond in reality to anionic vancancies, noted as black squares in Fig. 4.19. If they were occupied by oxide ions, the iron sites of these layers would become octahedral (dotted lines on the figure) and would lead to layers similar to the already existing perovskite layers, above and below the layer of tetrahedra. In this case, the formula would become $Ca_2(^{VI}M)_2O_6$ (or $CaMO_3$) with a charge +4 for the metal. $CaTiO_3$ effectively exists. It is a tilted perovskite.

Therefore, it can be said that $Ca_2Fe_2O_5$ is a defect perovskite structure, the defects being anionic. The developed formula $Ca_2(^{IV}Fe^{VI}Fe)O_5\square_1$ takes into better account both of two different coordinations of iron (tetrahedral ^{IV}Fe and octahedral ^{VI}Fe) and the existence of ordered anionic vacancies.

This sheds some light on the synthesis strategy of solid state chemists. When they discover a new structural type, their first reaction is to look at what other metals are compatible with this structure for extending this new structural family. For instance, if the example of brownmillerite is addressed, they will try to introduce other $3d$ transition metals in the place of Fe(III) and Al(III), keeping the same structure. If it works for iron- and cobalt-based solids, it is no longer true for manganese (Fig. 4.20).

Instead of the above succession of tetrahedral and octahedral layers, only a square pyramidal five-fold coordination is observed with manganese. These

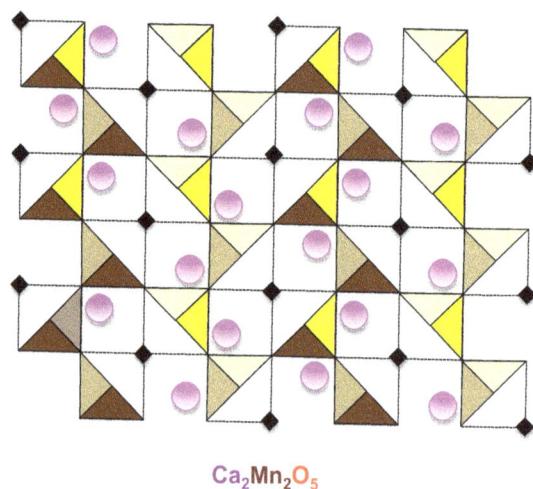

$Ca_2Mn_2O_5$

Fig. 4.20.   Idealized [001] projection of $Ca_2Mn_2O_5$ (Poeppelmeier, 1982) in which all the Mn^{3+} ions adopt a five-fold coordination (square-based pyramid). Vacancies appear as black squares.

pyramids, which share their five vertices, are in reality the visible part of octahedra (dotted lines in Fig. 4.20) becoming defective after the loss of one oxide ion.

Why such a difference compared to $Ca_2Fe_2O_5$? A possible explanation could be the size of the Mn^{3+} ion, the largest of the trivalent $3d$ cations ($r = 0.82$ Å), and therefore hardly compatible with a tetrahedral coordination, except during special conditions. This leads to another distribution of vacancies.

Of course, the classical oxidation of $Ca_2Mn_2O_5$ leads to the perovskite $CaMn^{4+}O_3$. However, for soft conditions of oxidation, an intermediary solid appears in the system $Ca_2Mn^{3+}_2O_5$–$CaMn^{4+}O_3$ with the formula $Ca_2Mn^{3+}Mn^{4+}O_{5.5}$, in which there are two valencies (III) and (IV) of manganese. Even if the structure is not known, the coexistence of octahedra and square pyramids can be anticipated.

The $Ca_2Mn_2O_5$–$Ca_2Fe_2O_5$ comparison illustrates the extreme influence of the size of the cation at the center of the polyhedron on the organization of the anionic vacancies within the three-dimensional framework, while keeping the same A cation. When A ions are heavy-metal ions (alone or in a mixture), the choice of the latter is also decisive for the resulting structure. The compound $BaLa_4Cu_5O_{13}$ (Michel *et al.*, 1985), a perovskite defective in anions, provides a nice example on this point (Fig. 4.21).

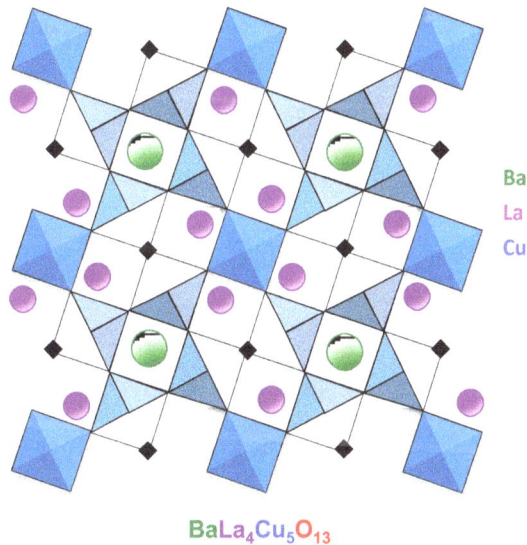

$BaLa_4Cu_5O_{13}$

Fig. 4.21. Idealized [001] projection of $BaLa_4Cu_5O_{13}$. Cu(II) and (III) ions are distributed in the octahedra and the square pyramids. The latter form tetramers which host Ba ions ($r_{Ba^{2+}} = 1.75$ Å) in twelve-fold sites similar to those of perovskite. Lanthanum ions ($r_{La^{3+}} = 1.18$ Å) occupy the tunnels with an octogonal section, as calcium ($r_{Ca^{2+}} = 1.20$ Å) in $Ca_2Mn_2O_5$ (Fig. 4.20). In the center of the latter tunnels, the anionic vacancies are represented as black squares.

The noticeable difference in the sizes of barium (1.75 Å) and lanthanum (1.30 Å) leads to an ordering between them. As for calcium in $Ca_2Mn_2O_5$, lanthanum occupies the octogonal tunnels, whereas the larger baryum, is in the dodecahedral sites similar to those of the ideal perovskite, the structure which is found when all the vacancies are occupied by the oxide ions.

Brownmillerite and its related solids provide a nice example of what intelligent crystal chemistry can mean to chemists. By the way, it also provides an unusual image of the oxidation phenomenon, which is represented in another way, different from the courses at the university. Moreover, this image allows to return in a rational manner to the synthesis of new products, and often to anticipate the adopted structure.

The second example for this family: $(La_{1-x}Sr_x)_8Fe_2Cu_6O_{20}$ (Fig. 4.22) and its homologue containing only copper (Genouel *et al.*, 1995) is very important. These two solids are only synthetic products and do not exist in nature. For the author, their structure is probably the most fascinating ever observed. Beside the lanthane/strontium mixture, which allows best dimension for the corresponding site, the coexistence of three different coordinations: IV, V and VI in the same framework is exceptional, even if the use of copper has its importance...

This structure leads to two major remarks: (i) the evidence of large elongated tunnels hosting the lanthanum and strontium ions and (ii) the existence of structural blocks $^VCu-^{IV}Cu-^VCu$ (the exponent relates to the coordination of copper)

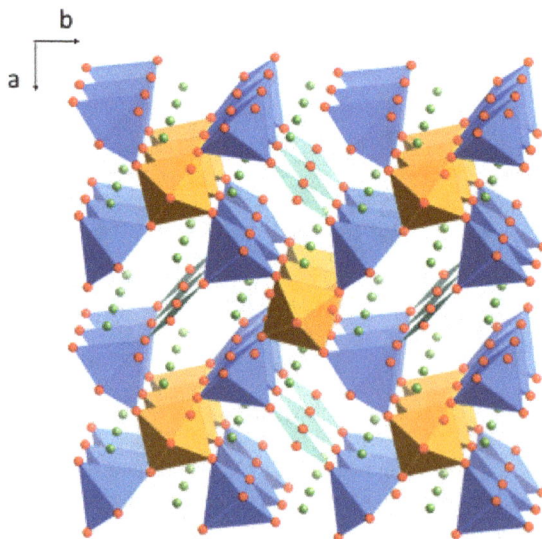

Fig. 4.22.   Perspective view of $(La,Sr)_8Fe_2Cu_6O_{20}$ (Cu: blue (deep for five-fold, light for square planes); iron: orange; La,Sr: green; oxygen: red).

identical to those existing in the already described $YBa_2Cu_3O_7$. In spite of the identity of these blocks in the two structures, their organization is clearly different. Parallel in YBaCuO, the blocks arrange in an orthogonal way in the above solid (Fig. 4.19(right)). A question immediately arises from this difference: does the change of orientation have an influence on the possibility of vacancies, as was the case for the superconductor and its ability to vary its composition from O_6 to O_8, once the vacancies are filled by oxide ions, with a correlative change of the valency of copper?

The reply to this question implies a more quantitative comparison, first on the formula of the motif using the rules edicted before. The formulation of the copper trimer in both structures is Cu_3O_7 $[(CuO_{5/2})_2(CuO_{4/2})_1]$ and that of the iron octahe-dral chains is FeO_3 $[(FeO_{6/2})]$. In YBaCuO, the trimers are catenated (**catenation**: the capability of a group to link with similar others) and present for the framework the $[Cu_3O_7]\infty$ formula, inside which Y and Ba are hosted; in the (La,Sr) com-pound, the trimers are linked through vertices to the files of octahedra. The deduced formula $[(Cu_3O_7)(FeO_3)]_\infty$ corresponds well to the chemical formula. Nevertheless, at variance to what happened for YBaCuO, the La/Sr species occupy only a single type of site, but what site? When looking at the two structures, bar-yum in YBaCuO and the mixture La/Sr in the other solid **locally** occupy *the same position* in regard to the trimer in both structures, with similar X coordinations. It is therefore in the different orientations of the blocks that the relation must be investigated (Fig. 4.19).

A new feature rapidly appears: in $YBa_2Cu_3O_7$, cubic sites exist, occupied by yttrium. They are also found in $(La_{1-x}Sr_x)_8Fe_2Cu_6O_{20}$, but this time not occupied. From this observation, does it mean that the last compound is defective? The answer is not so easy because it can be dependent on different local situations.

Indeed, in YBaCuO, the cube is formed by the square faces of two tetragonal pyramids belonging to two different blocks Cu_3O_7. These faces are separated by a distance 2.864 Å, which correspond to the tangency of two oxide ions, allowing the insertion of Y. On the contrary, in $(La_{1-x}Sr_x)_8Fe_2Cu_6O_{20}$, the opposite faces which form the cube show, for one, the square in the middle of a Cu_3O_7 entity, and, for the other, the square basal plane of another Cu_3O_7 group. The O–O distance between the opposite faces is larger (3.365 Å), due to the steric hindrance of the pyramids. The site is effectively empty but does it correspond to a fillable vacancy? Clearly not. Owing to the cubic environment of oxide ions, the eventual filling would be performed by cations, but the electroneutrality rule would need to simul-taneously introduce other oxide ions which that do not have sufficient place in the existing anionic subnetwork. This is not the case for YBaCuO: there is enough place between the two square planes of two face-to-face blocks, which explains that the composition of YBaCuO can reach a formulation in O_8. Therefore it seems

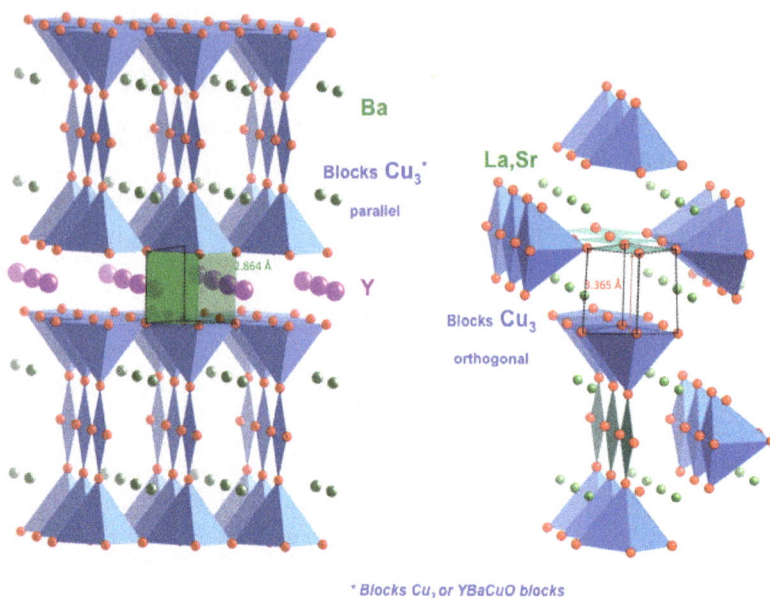

Fig. 4.23. Comparison of the local situations around the Cu_3O_7 trimers in the two structures. The origin of the cubic sites, filled in YBaCuO, empty in the other, is specific in both cases.

that the only **apparent** variable parameter would be the La/Sr ratio inside the ten-fold coordinated site.

This last example concludes this chapter. Meanwhile, the next chapter follows with its new ways for looking at a structure. Here, the description of assemblies was done, not starting from the different modes of connection of polyhedra, considered as initial bricks, but from more complex blocks. Once identified, the latter not only allows a more synthetic, simplifying and memorizable version of the structure (which is the ultimate aim of an analytical crystal chemist!), but also serves as a springboard for imagining a *creative* chemistry from the analyzed descriptions. This will be seen in Chapter 7, but as before, other ways for analyzing structures are useful and more and more topological. They will be described in the following brief chapter.

Chapter 5

Armatures, Gilders and Bricks!... Other Ways of Reading a Structure

Less is more ....
Mies van der Rohe
American Architect (1886–1969)

This choice of title of course implies architecture and the construction of buildings... It is true but, in this chapter, the same strategy will be used but, this time, at the atomic scale for the scientist who wants to rationally create a new structure. As for the architect, he must first draw sketches on the paper, then determine the inner superstructures that will serve as a skeleton for the future molecular building, and the gilders linking them. It is only after that, walls will be constructed with bricks, sheets for a segmentation of the space... Decoration occurs later...

In this book, what have we realized up to now? Just an interest for the bricks and their assembly. The first bricks were the polyhedra, and interest was focused on their connectivity and the resulting dimensionalities. With the progress of knowledge, the diverse representations, more and more complex, incited the scientist to describe the complexity with new tools aiming for an easier understanding of this complexity. It was an analytical training of the eye and a basis for a different approach to creativity. It was a necessary — but not sufficient — step. But now, what about the superstructures? It is on this point that this chapter emphasizes.

Fig. 5.1. The steps of a macroscopic building site.

A.F. Wells, B.G. Hyde and M. O'Keeffe, already cited, were the pioneers of this approach, but their vocabulary was not the vocabulary of architects. They call superstructures nets; for gilders, they prefer rods... These words will appear in the following.

5.1 Nets

They concern only one part of the entire structure, but this part will influence the construction. This part can be anionic, neutral or cationic, depending on the interest. The aim is to first isolate this pertinent part and to exploit its corresponding planar subnetwork in relation with the question it can solve and the transformations it can undergo under the action of an external stimulus.

This paragraph is far from being purely academic. It allows for example to have a structural vision of the adsorption of species on a surface (an important aspect of catalytic phenomena) just by looking at the atomic scale — not only the organization of this surface, but also the possible sites where the molecules can reasonably be adsorbed (Fig. 5.2).

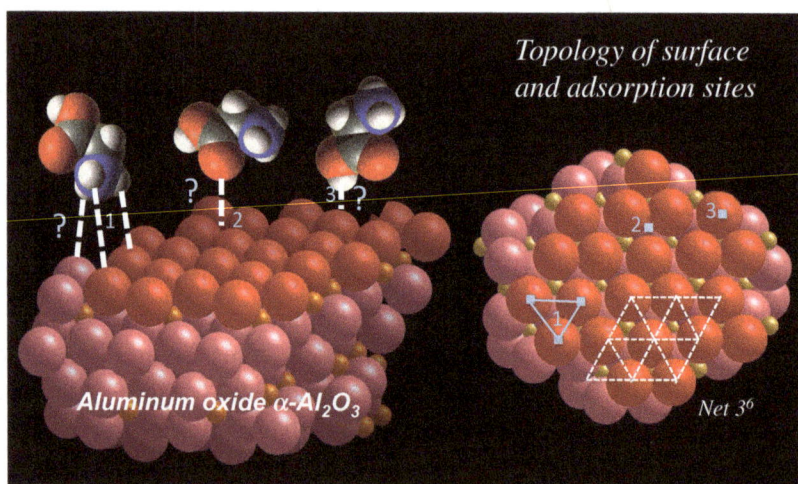

Fig. 5.2.   Perspective (left) and projection (right) views of the three possibilities of adsorption of a molecule of glycine on the (001) plane of a perfectly planar surface of aluminum oxide Al_2O_3 in its α form (corundum). This surface consists of oxide ions forming a 3^6 net (dotted lines). Depending on the orientation of the molecule when the contact occurs, either the hydrogen atoms of the ammonium group, or the oxygen atoms of glycine are considered. Three situations can occur: (i) hydrogens of the ammonium group can fix on a triangle of oxide ions (case 1); when the carboxylate function is adsorbed, two other possibilities: (ii) the oxygen of the double bond C=O of the carboxylate function arrives above the center of a triangle of oxide ions (case 2), or the hydrogen of this function interacts with only one oxide ion of the surface (case 3). The solution is experimentally provided by spectroscopies (oxygen: red; carbon: grey; hydrogen: white; nitrogen: blue).

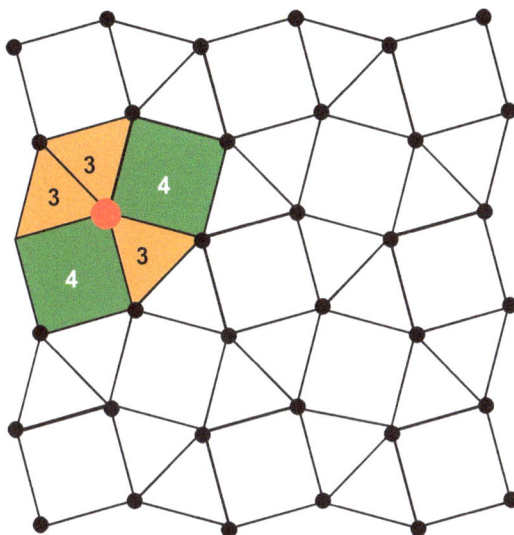

Fig. 5.3. Explanation of the Schläfli notation for a $3^2.4.3.4$ net.

The Schläfli notation was defined in Chapter 1 for a symbolic description of Platonic and Archimedian polyhedra. It also applies to plane nets because, in two dimensions, polyhedra are replaced by polygons, and Platonic and Archimedian polygons also exist. For example, the notation of the Archimedian network represented in Fig. 5.3 is $3^2.4.3.4$.

There are three Platonic polygons (identical regular polygons) (Fig. 5.4), and eight Archimedian ones (regular polygons, but of several types) (Fig. 5.5).

These simplified representations contain much useful information: (i) help for understanding the structural modifications by considering these nets as flexible (each vertex is assimilated to a kneecap); (ii) render reliably similar topologies existing in different chemical families that, without this concept, would be almost impossible to compare, either their formula or their properties.

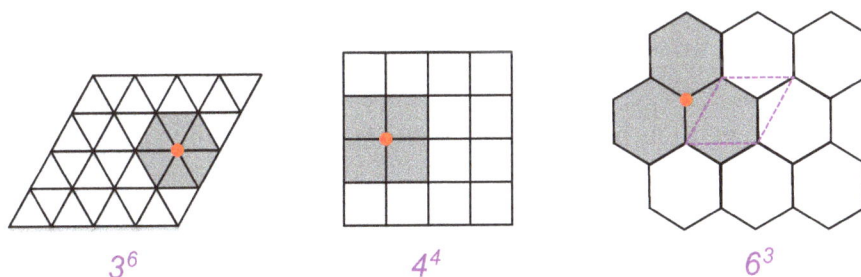

3^6 4^4 6^3

Fig. 5.4. The three Platonic polygons. The red points explain the Schläfli notation. Note that 3^6 and 6^3 are dual forms; 4^4 is self-dual.

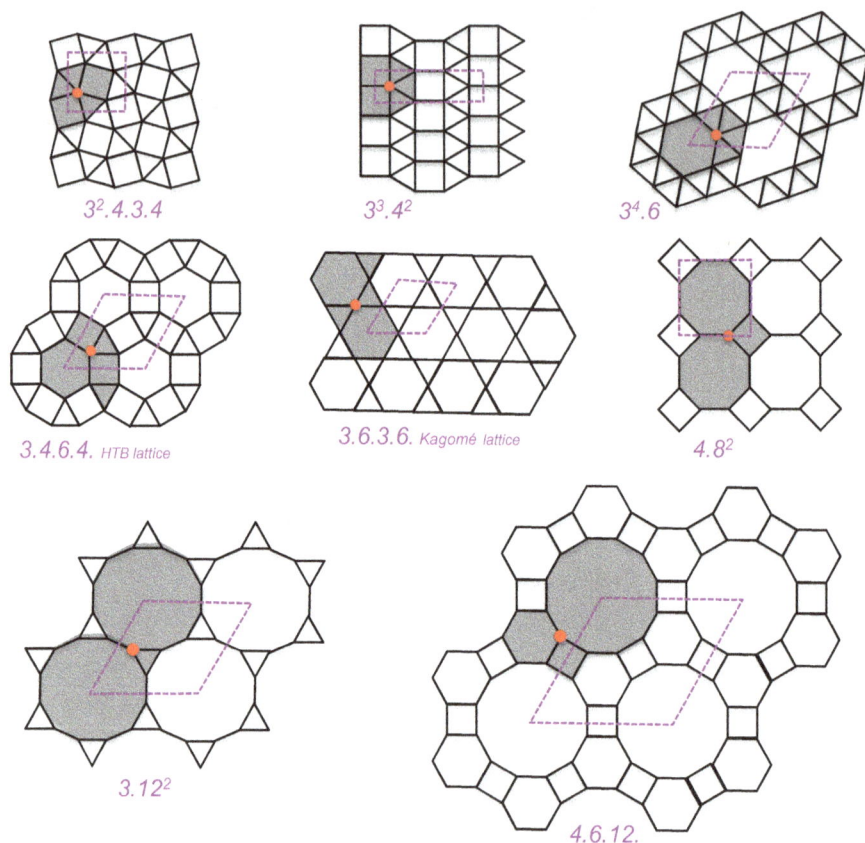

$3^2.4.3.4$ $3^3.4^2$ $3^4.6$

$3.4.6.4.$ HTB lattice $3.6.3.6.$ Kagomé lattice 4.8^2

3.12^2

$4.6.12.$

Fig. 5.5.   The eight Archimedian polygons. Cells are indicated as dotted lines.

5.1.1 *Flexible nets*

This notion was briefly discussed in Chapter 4 when the crystallographic shears with their cooperative slidings were presented.

In a first approach, consider four particles forming a dense triangular net 3^6 [Fig. 5.6(a)]. Opposite **glidings** of two horizontal rows transform two triangles into a square, while eliminating the bond corresponding to the small diagonal of the initial lozenge. Applied to Archimedian nets, this operation allows to observe a relation between three of them, facilitated by specific glidings [Fig. 5.6(b)].

Coming back to chemistry, these planar arrangements are the bases of dense stackings, described in Chapter 1. A triangle is the face of an octahedron and the square of a face of a cube. It is on these faces that a cation is fixed to further create a polyhedron (Fig. 5.7). The gliding operation therefore justifies the reason for the change of coordination observed in some solids.

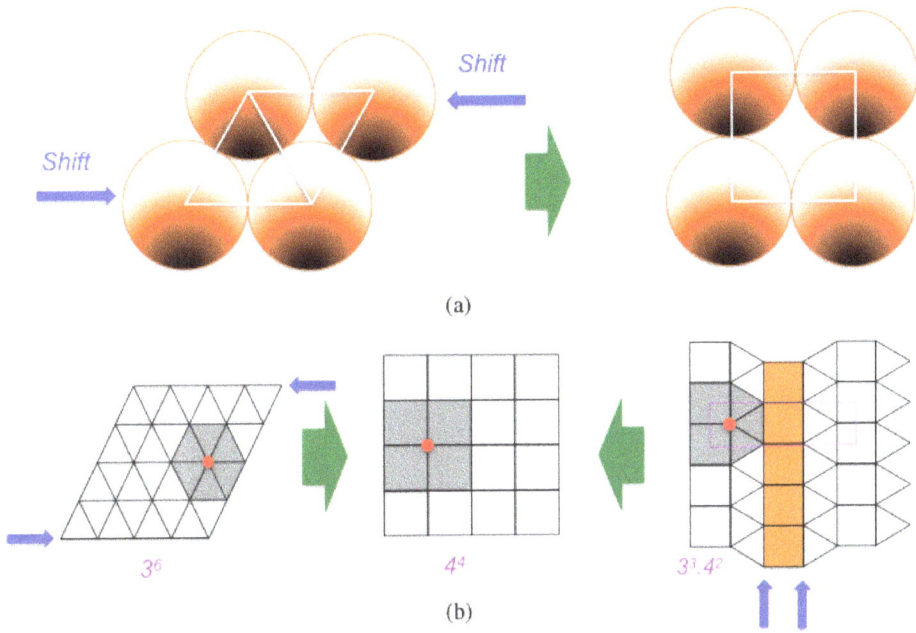

(a)

(b)

3^6 4^4 $3^3.4^2$

Fig. 5.6. (a) Passage triangle → square by antagonistic glidings; (b) application to Archimedian nets and their relations explained by this operation; the gliding directions are indicated by the blue arrow.

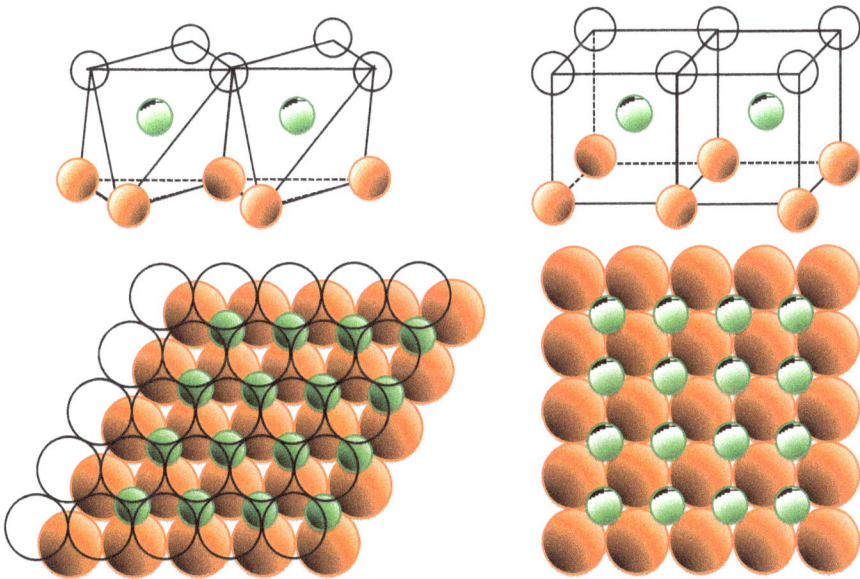

Fig. 5.7. Cooperative gliding of anions (in orange) and the associated changes of coordination, from octahedral to cubic (cations appear in green and the black empty circles outline the anions of the above layer for a reconstitution of the polyhedron).

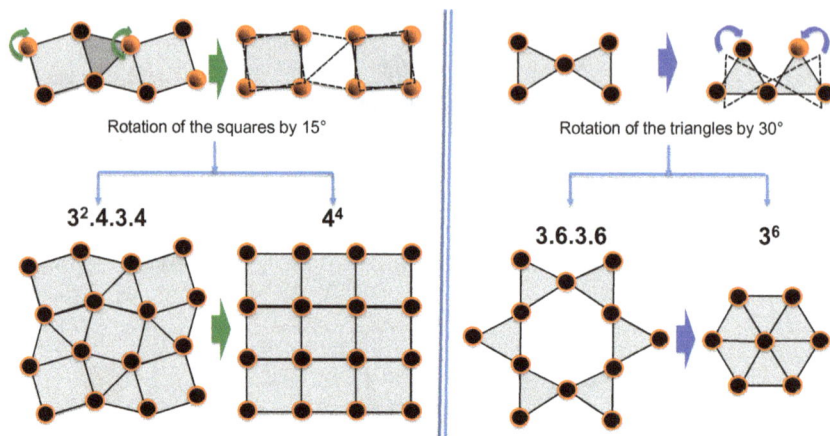

Fig. 5.8. Cooperative rotation of anions (in orange) and the associated changes of coordination. Unexpectedly, the $3^2.4.3.4$ net was encountered during the study of tilts of octahedra. The 3.6.3.6 net is often encountered in the structures of silicates based on tetrahedra.

However, is the gliding the only operator for passing from one net to the other? Obviously not! Another of great importance can be seen in the following: the **rotation** that also implies weak displacements of the atoms, but can lead to drastic modification of the topology. Here also, two examples (Fig. 5.8).

5.1.2 *Decorated nets*

The above examples connect triangles, squares and hexagons, all polyhedra which are able to completely cover the plane. What happens now with regular pentagons? Mathematicians have proved since a long time that only pentagons are unable to completely cover the plane. This would lead to 4% of the plane being uncovered. It is also true for nets but, accepting some slight distortions of the pentagons, Nature — less strict than mathematicians! — solved the problem and nets containing pentagons very often exist (Fig. 5.9).

The net of mercury atoms in Mn_2Hg_5, even not Archimedian, has considerable importance in the Magnéli's bronzes family, these berthollides mentioned during the presentation of non-stoichiometry.

This remark introduces the second advantage of planar nets: establish a relation between structures which, apparently, have nothing in common, neither in their formula not in their bonding nature (intermetallics for Mn_2Hg_5 and ionic oxide for the Magnéli bronze). This is again another (indirect) proof that it is the minimization of the lattice energy of the framework which governs the structure of solids, whatever the nature of their bonds.

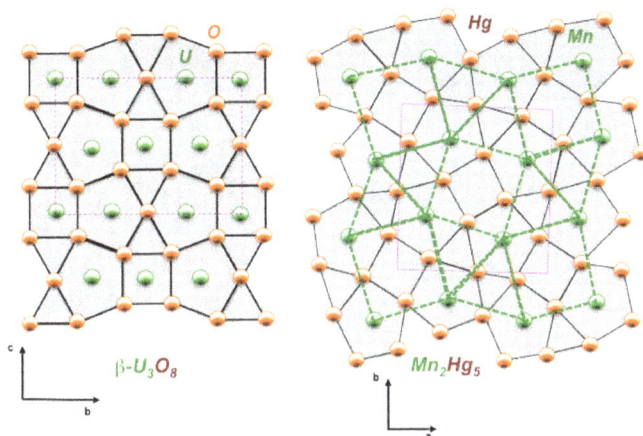

Fig. 5.9. Two examples of nets with pentagons (in orange). This time, each node of the net connects four polygons. In each example, the green circles correspond to the plane above the basal pentagonal net. Note that, in Mn_2Hg_5, the net of the atoms of manganese (green dotted line) is the $3^2.4.3.4$ net already seen in Archimedian nets (Fig. 5.5), whereas the Hg net (3.5.4.5) is not Archimedian.

What is the relation between Mn_2Hg_5 and the tetragonal bronze TTB of Magnéli K_xWO_3 ($x \approx 0.5$) (Magnéli, 1949b)? It becomes obvious when considering the two projections (Fig. 5.10). The Hg atoms are now replaced by the WO_6 octahedra through a possible metal to polyhedron, "decoration" of the Hg site by an octahedron, here.

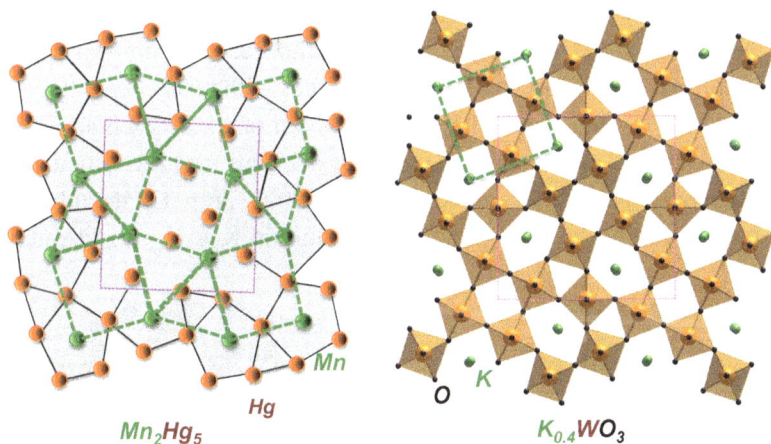

Fig. 5.10. Comparison of the topologies of Mn_2Hg_5 and the TTB bronze of Magnéli ($K_{0.4}WO_3$ or $K_2W_5O_{15}$). The nets of Mn in Mn_2Hg_5 and K in $K_{0.4}WO_3$ are identical. This is the same for the nets of Hg and WO_6 octahedra. To easily compare the two structures, focus on the green squares in both structures. Note that if $x = 0.6$ in K_xWO_3, the empty squares become occupied, which is not the case for Mn_2Hg_5.

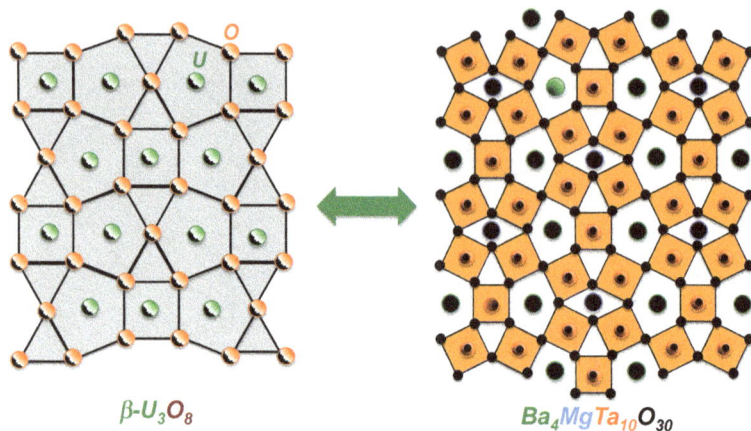

β-U_3O_8 $Ba_4MgTa_{10}O_{30}$

Fig. 5.11.   Comparison of the topologies of β–U_3O_8 and of the bronzoid $Ba_4MgTa_{10}O_{30}$ (since iso-structural with tungsten bronze). The oxygen net in β-U_3O_8 and that of tantalum octahedra are identi-cal; it is the same for the other nets: the two square and pentagonal nets of uranium are respectively occupied by magnesium and barium in $Ba_4MgTa_{10}O_{30}$. In the last case, the differentiation Mg–Ba relates to the respective sizes of small Mg^{2+} in square sites and large Ba^{2+} in the larger pentagons.

This remark is also valid for β–U_3O_8, compared to the complex oxide $Ba_4MgTa_{10}O_{30}$ (Fig. 5.11).

This time, regarding "decoration", it is the anion which is decorated by an octahedron.

When the α form U_3O_8 is considered (Fig. 5.12), there is another situation that avoids too quick conclusions about the "decorated" species. The comparison occurs this time with the vanadate $K_3V_5O_{14}$. Indeed, up to now, the "decoration" was per-formed by octahedra, but it is not a general rule. Indeed, in the comparison α–U_3O_8–$K_3V_5O_{14}$, uranium cations occupy two types of sites with respectively three and four uranium neighbors. This means that the decoration of the oxygen sites of α–U_3O_8 by polyhedra will reflect the original two sites and will lead this time to two types of decorating polyhedra: two octahedra and one tetrahedron. On the other hand, the nets of uranium atoms in α–U_3O_8 and of potassium in $K_3V_5O_{14}$ are identical.

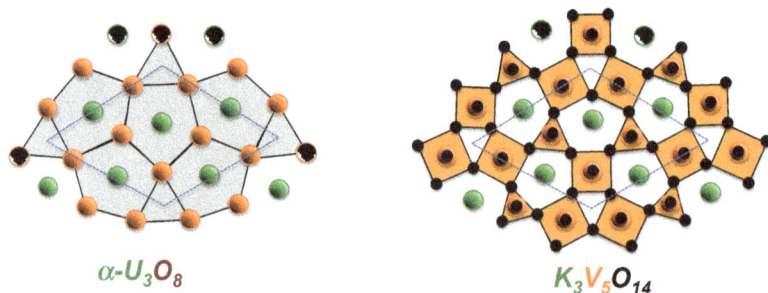

α-U_3O_8 $K_3V_5O_{14}$

Fig. 5.12.   Compared topologies of α–U_3O_8 and $K_3V_5O_{14}$. The codes of color are the same as in Fig. 5.11.

Such correlations are numerous between intermetallics and oxides and other chalcogenides, therefore showing the richness of the solid state chemistry. They build an unexpected but rich bridge between apparently completely different chemical families. All of them have in common a creation which always minimizes their lattice energy, whatever the nature of the elements and their bondings. Structures have a thermodynamic origin. Their geometry is just the emerging part of the iceberg...

This paragraph has also introduced the concept of "decoration" for establishing relations between these different structures. As already mentioned, it is not just an academic exercise of analysis. Chapter 7 will demonstrate that, far beyond the analytical description of structures, a powerful tool can be provided for the rational creation of new surprising and interesting edifices.

5.2 Rod Descriptions of Structures (O'Keeffe, 1992)

They are not as powerful as the descriptions in terms of nets. However, they offer a tool for simplification allowing the easier memorization of structures. The approach is this time completely inverse of the latter. Instead of scrutinizing the local arrangements of atoms, the observer, on the contrary, will consider the situation with more detachment and be more interested by associations of polyhedra instead of the polyhedra themselves. It is particularly true when polyhedra form chains in several directions of the space.

Take for instance the now well-known rutile structure. It was already described either in terms of dense packings or by associations of octahedra. But both were centered on the octahedron, as the interstice in dense packings or as a brick of construction connected to others. The description in terms of rods forgets the octahedron and focuses more on the connection of larger assemblies, independently of the nature of the polyhedra forming this assembly (the infinite chain of edge-shared octahedra in rutile (Fig. 5.13). This better proves the armatures of the

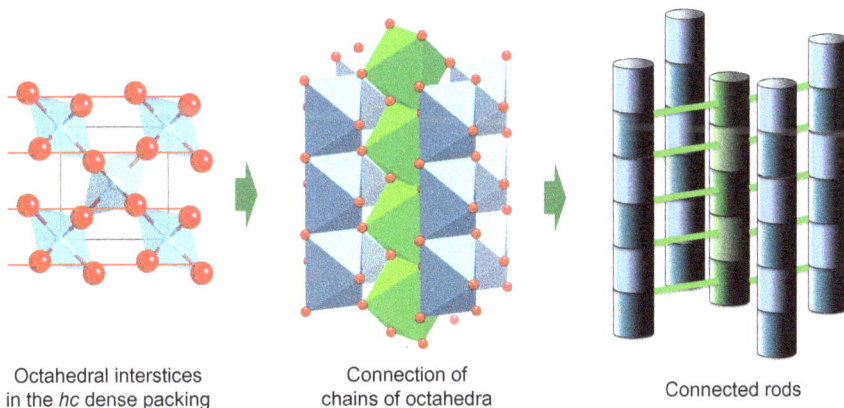

Octahedral interstices in the *hc* dense packing

Connection of chains of octahedra

Connected rods

Fig. 5.13. Evolution of the description of the rutile structure (case of parallel chains).

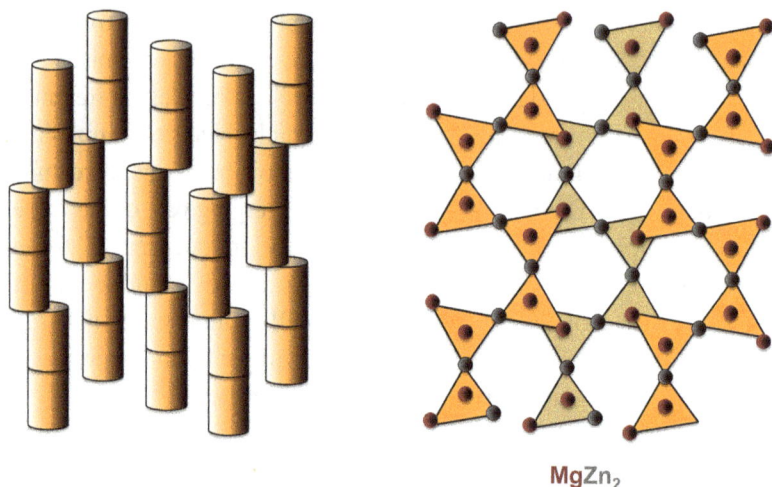

MgZn₂

Fig. 5.14. The rod description of the intermetallic MgZn$_2$ compared to the usual one using only associations of polyhedra. In the latter description, MgZn$_2$ is built from infinite zig-zag chains of trigonal bipyramids sharing both vertices and edges.

considered structure. These assemblies can be considered as independent or will take into account their specific connections, as represented in Fig. 5.13.

The simplicity of this first example has the advantage to present a more synthetic vision, but it remains rather trivial. One can refine the representation in the case of zig-zag chains which appear for instance in MgZn$_2$ (Fig. 5.14).

This rod representation is more useful when chains are crossed in different manners, forming primitive or centered networks along one, two or three directions. This allows again to make closer structures apparently different, like an architect who imagines different buildings using the same armatures. Some examples follow.

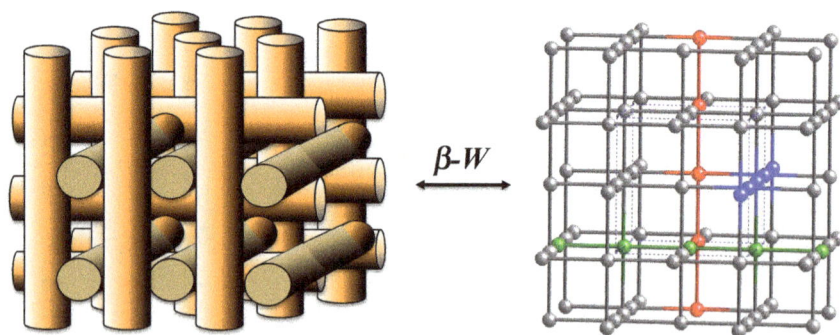

β-W

Fig. 5.15. Infinite rods crossed in the three directions of the space. This representation can describe the structure of the β form of metallic tungsten. In each direction, the rods form a primitive sublattice.

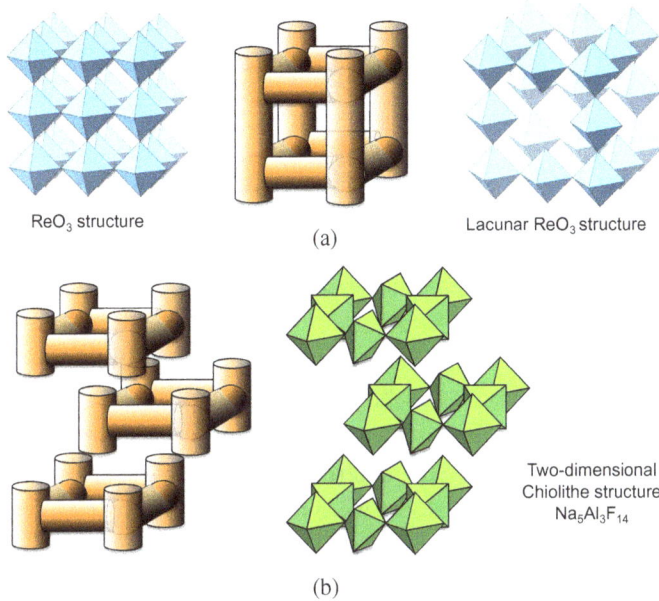

ReO$_3$ structure   Lacunar ReO$_3$ structure

(a)

Two-dimensional
Chiolithe structure
Na$_5$Al$_3$F$_{14}$

(b)

Fig. 5.16. Examples of interpenetrated rods in three or two dimensions with their examples.

Crossed rods in Fig. 5.15, can also be interpenetrated in two or three dimensions (Fig. 5.16).

Figure 5.17 illustrates a frequent case: the face centered disposition of orthogonal networks.

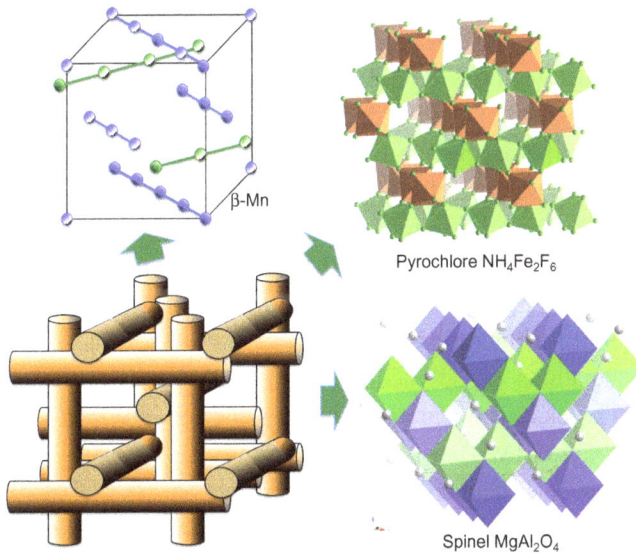

β-Mn

Pyrochlore NH$_4$Fe$_2$F$_6$

Spinel MgAl$_2$O$_4$

Fig. 5.17. Face-centered disposition of a three-dimensional network of crossed rods.

This variety of possible dispositions serve as a common denominator between architectural structures apparently without any common point, whatever the point: formula, chemical or structural.

5.3 Building Units and Descriptions of Structures (Férey, 2000)

This approach is rather new, and starts from the strategy already employed for rods: the observer is far from the object.

Before going further, it may be interesting to understand why, during such a long time, every description was based on the polyhedron. It probably comes from the fact that crystal chemistry was born from crystallography, this science which, from the study of a crystal, determined symmetries, crystallographic sites with their coordinates, where atoms were located. Initially, it was therefore normal to focus on the formular object for which, since Pauling, the metal was the center. This quasi-molecular vision remained predominant up to the time when scientists began to look at the environments at medium, then long distance. From this point of view, high resolution electron microscopy became and is more and more an essential technique for understanding the long distance organization of the solid, with its periodicities, its defects, giving finally the true image of the **real** solid.

Post remark, and even if crystal chemistry can be performed without the knowledge of the formula and the symmetry, a question remains: is the use of the polyhedron an obligatory passage during the description? Some of us have recently proved that it was not obligatory, by initiating a new concept: **the scale chemistry concept**, which will be developed in detail in Chapter 7. The question is the following: instead of considering the polyhedron as the basic brick (as done in Chapter 4), is it possible to find larger associations of polyhedra which, respecting the periodicity of the solid, could play the same role? The answer is positive, with some associated important advantages: (i) a better relation between the structure, the formula and the number of motifs per cell; (ii) a simpler and more abstract representation of the solid and, last but not least (iii) the creation of a powerful tool for the rational creation of new interesting solids.

These associations of polyhedra will be called not-so-differently building bricks, buiding blocks, secondary building units (acronym SBU), molecular building blocks (acronym MMB) or simpler bricks.

With this concept, chemists and architects are very close. Chemistry, as architecture, is a construction. With time, architects have imagined and built from different kinds of bricks many types of buildings, (Fig. 5.18), guided by their sense

Fig. 5.18. Some architectures based on various bricks, function of the periods, the localizations, the types of climate, aesthetic feelings and the technical expertise of the architects who conceived them.

of aesthetics, constrained by the necessary stability of their resulting building. Chemists have also created many kinds of solids, this time at the molecular level, with the constraints of lattice energy and minima of free energy. Only the size of the bricks changes, from macroscopic to microscopic…

To briefly illustrate this new concept, take some examples of simple structures already described.

The chiolite structure [Fig. 5.19(a)] was described above in terms of both octahedra and rods. The same can be done with bricks [Fig. 5.19(b)]. It can indeed be decomposed into trimers of octahedra, whose connections by vertices regenerate the macroscopic two-dimensional solid. In this operation, the octahedron is replaced by a new building block which is the trimer of these polyhedra. This provides some advantages: (i) first for the formula: the choice of the SBU agrees now with the chemical formula of the compound; the developed formula of the trimer, which takes into account the connection modes of the central and satellite octahedra, is $[AlF_{4/2}F_{2/1}][AlF_{4/1}F_{2/2}]_2$ therefore Al_3F_{14}; Na^+ ions are considered as species inserted between the layers of the structure; (ii) simplification of the lecture of the structure: it is now a centered association of bricks [Fig. 5.19(c)] as we did in Chapter 1 for a centered cubic structure.

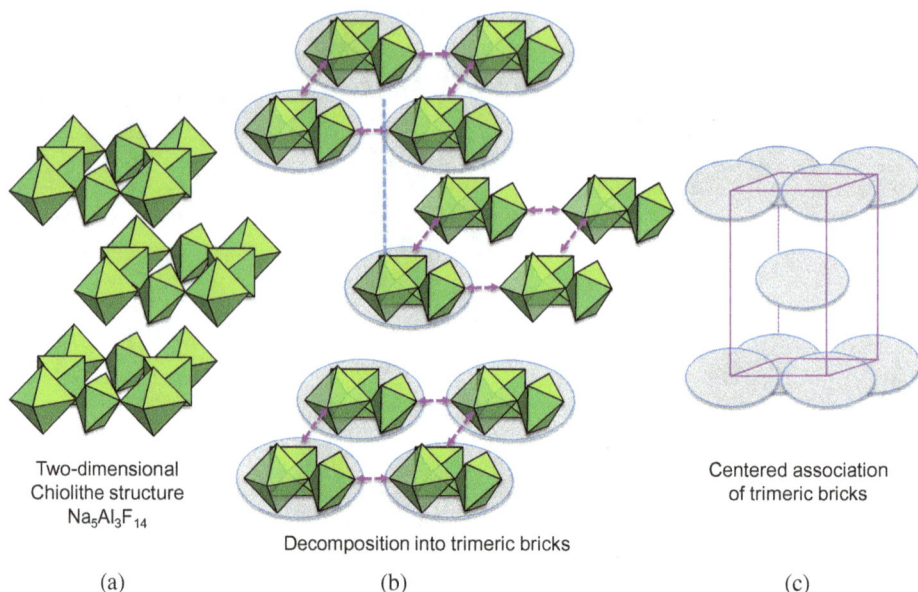

Two-dimensional
Chiolithe structure
$Na_5Al_3F_{14}$

Decomposition into trimeric bricks

Centered association
of trimeric bricks

(a)   (b)   (c)

Fig. 5.19.   The evolution of the description of chiolithe, from layers of corner-shared octahedra to a centered association of trimeric bricks.

Once this concept is understood, the comparison between the $BaNb_2O_6$ and $CaTa_2O_6$ structures (Fig. 5.20) becomes trivial. Even if their dispositions differ, both are built up from dimeric bricks of edge-shared octahedra. The new brick takes into account the formula with two metals.

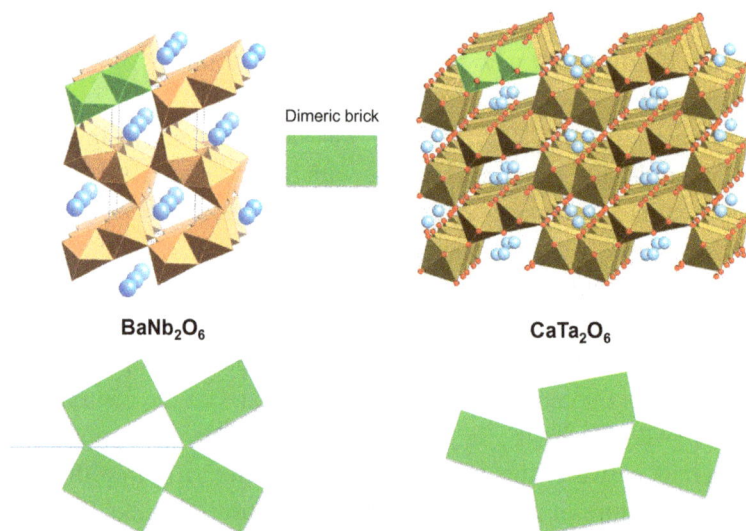

Dimeric brick

$BaNb_2O_6$ $CaTa_2O_6$

Fig. 5.20.   Nature and organization of the dimeric bricks in $BaNb_2O_6$ and $CaTa_2O_6$.

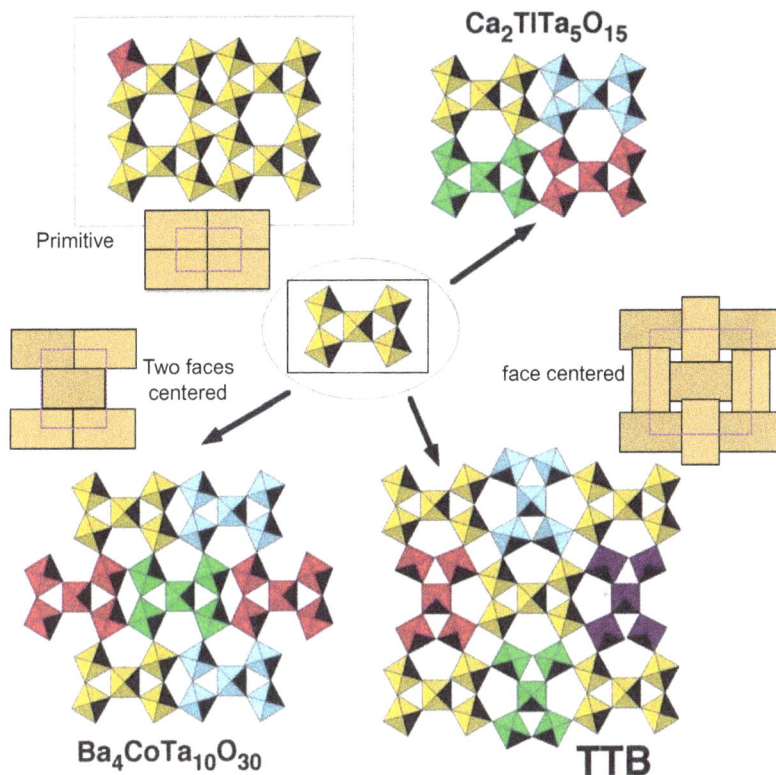

Fig. 5.21. Some structures based on the pentamer M_5X_{15}. The different colors are just here for highlighting the different pentamers involved in the structures. The cations in the different tunnels are omitted for sake of clarity. The apparently different formulae are in reality multiples of a A_xMX_3 skeleton. If the TTB bronze of Magnéli is written K_xWO_3, the others could be also written as $(Ca_2Tl)_{0.2}TaO_3$ and $(Ba_4Co)_{0.1}TaO_3$, respectively.

The building blocks can be larger, as in the case of the tetragonal bronze TTB of Magnéli K_xMX_3 ($x \approx 0.5$). It was previously described as rotated perovskite blocks, but it can also be considered as those formed from the association of pentamers of vertex-shared octahedra in a bow-tie shape (Fig. 5.21).

The description of structures in terms of building blocks allows easier relations between different structures. Not only does TTB correspond to a two-face association of pentamers, but this simplified description allows to relate it to different structures based on this pentamer, with diverse organizations and orientations, depending on the ratio between the sizes of the two cations hosted in the tunnels. One pass from $Ca_2TlTa_5O_{15}$ to $Ba_4CoTa_{10}O_{30}$ is simply the half-cell parameter gliding over two rows of pentamers, and to TTB by a 90° rotation of one-half of the

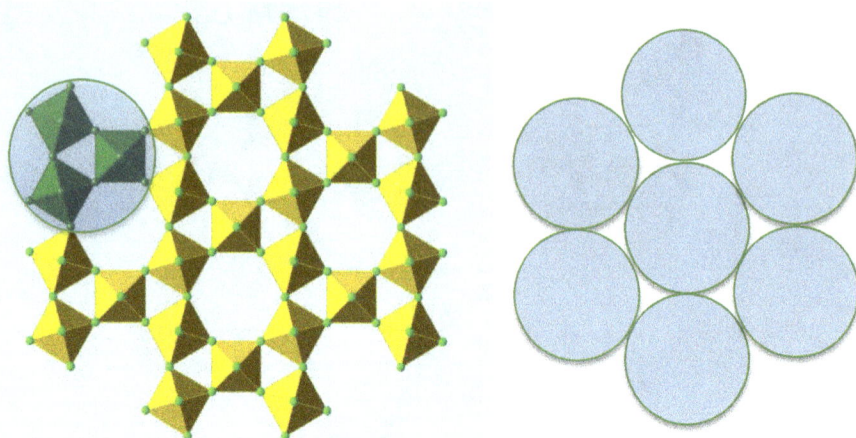

Fig. 5.22.   The structure of the hexagonal bronze of tungsten (HTB) described from trimers of corner-shared of octahedra. One trimer appears on the left in green in the circle These circles form a dense hexagonal plane.

pentamers. These different operations modify the shape of the majority of the tunnels (hexagonal in $Ca_2TlTa_5O_{15}$; pentagonal in $Ba_4CoTa_{10}O_{30}$; mixed pentagon-squares in the TTB. Whatever the structure, the triangular tunnels are maintained.

The hexagonal bronze of Magnéli HTB, which also results from the connection of M_3X_9 trimers of corner-sharing octahedra, can be treated in the same way (Fig. 5.22). Despite being based on trimers, both the bronze and the chiolite have different formulae (M_3X_9 for the bronze, M_3X_{14} for chiolite).

Further simplifying, the trimers in a plane can be assimilated to spheres. In this case, the resulting organization is equivalent to the dense plane described in Chapter 3.

Note that, at variance to what happens in the hexagonal dense packing, the dense planes of HTB just stack one over the other for regenerating the whole structure, with no tilting of octahedra of the trimer from one plane to the other.

These HTB layers can also exist when the trimers are tilted in a structure. It occurs in the 2D solid $Cs_4Cr_5F_{19}$ [Figs. 5.23(a)–5.23(c)]. An antiphase rocking of the axial anions of the trimer renders them tangent. They form a triangle with the same dimensions as the face of an octahedron. This allows the possibility of grafting a supplementary octahedron form on the layer, leading to the formation of a tetrahedral Cr_4F_{18} tetramer of octahedra [Fig. 5.23(b)], instead of the initial trimer. The connection [Fig. 5.23(c)] of these super-tetrahedra by vertices provides the HTB layers on which a supplementary octahedron is grafted. The organization of

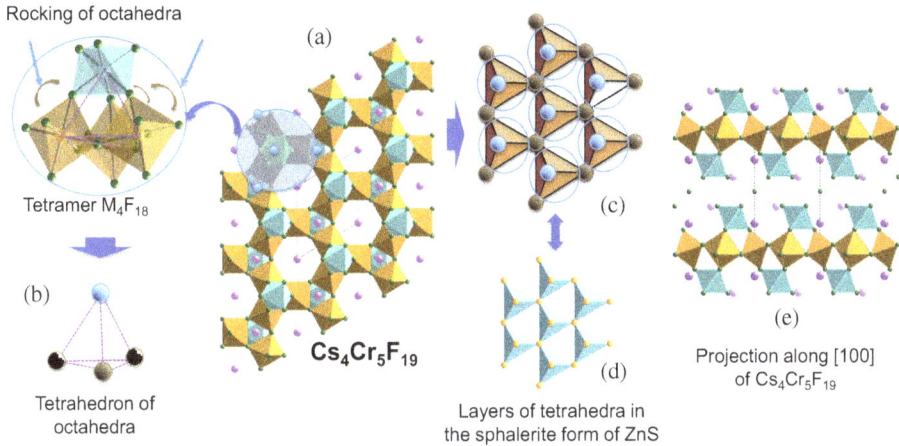

Rocking of octahedra

Tetramer M_4F_{18}

(a)

(b)

(c)

(d)

(e)

Tetrahedron of octahedra

$Cs_4Cr_5F_{19}$

Layers of tetrahedra in the sphalerite form of ZnS

Projection along [100] of $Cs_4Cr_5F_{19}$

Fig. 5.23. (a) [001 projection] of $Cs_4Cr_5F_{19}$; the tetramer of octahedra and its cationic skeleton, inside the circle, appear in perspective in (b); (c,d) homothetic topologies of supertetrahedra layers in $Cs_4Cr_5F_{19}$ (c) and in ZnS (d); alternation of layers and Cs...F chains in $Cs_4Cr_5F_{19}$ ([100] projection).

these layers is reminiscent of those existing in the two forms of ZnS [Figs. 5.23(c)]. Between such layers, the last fluoride ion and Cs^+ form …Cs^+…F^-… chains.

The same type of description based on supertetrahedra [Figs. 5.24(c)–5.24(e)] can be applied to the pyrochlore structure [Fig. 5.24(a)] and complement its previous rod description [Fig. 5.24(b)].

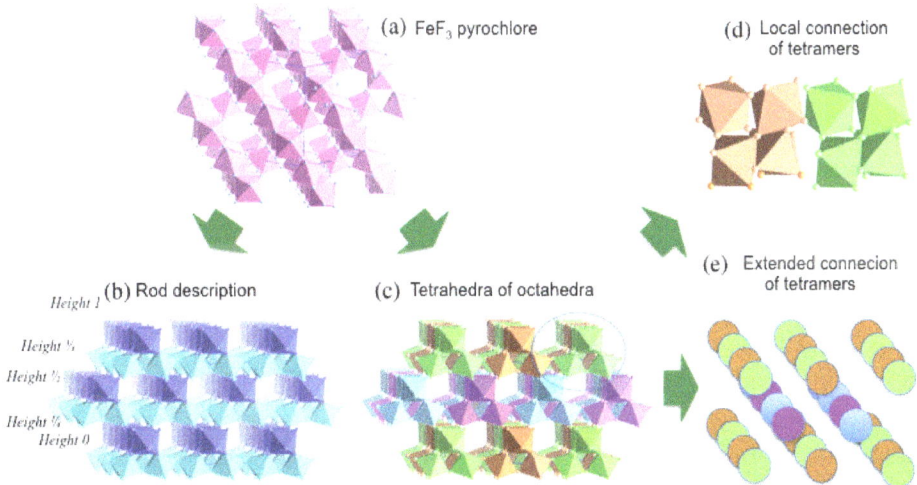

(a) FeF_3 pyrochlore

(d) Local connection of tetramers

(b) Rod description

Height 1
Height ¾
Height ½
Height ¼
Height 0

(c) Tetrahedra of octahedra

(e) Extended connecion of tetramers

Fig. 5.24. Diverse descriptions of the pyrochlore form of FeF_3.

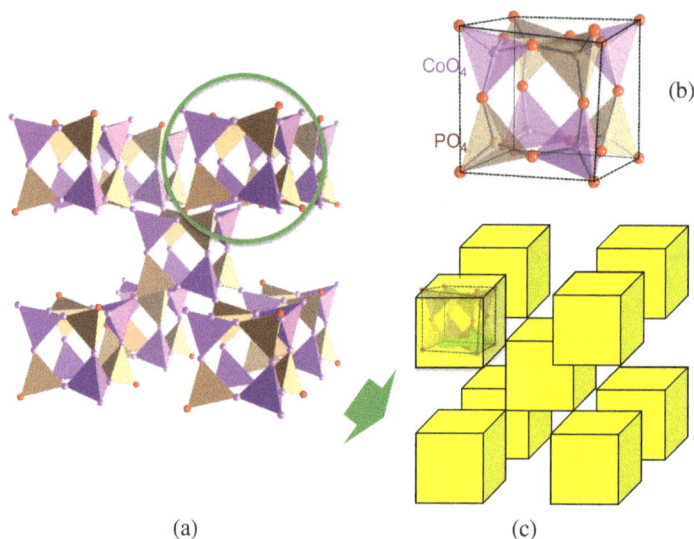

Fig. 5.25. (a) Perspective view of the cobalt phosphate; (b) the octamer of tetrahedra; (c) the centered cubic arrangement of the corner-shared octamers in $CoPO_4$.

The tetrameric building blocks, which correspond to tetrahedra of corner-shared octahedra, are linked together via vertices and form chains in the three directions of the space [Fig. 5.24(e)]. Note that, within the tetramers, a cationic ordering can occur. It was already mentioned with the Fe^{2+}–Fe^{3+} ordering in $NH_4Fe_2F_6$ (Fig. 5.17).

Cubes of tetrahedra (noted D4R for Double 4-Rings) will be the last example of simplification of structures using the building block concept. This often occurs in the family of porous zeolites. A nice example is provided by a cobalt phosphate (Fig. 5.25).

The building block is an octamer of corner-shared tetrahedra alternatively occupied by the phosphorous ions of the phosphate groups and by Co^{3+} ions in tetrahedral coordination. The outer oxygen atoms of these octamers define cubes which will serve as building blocks for describing the structure of $CoPO_4$ as a centered cubic corner sharing arrangement.

5.4 The "Spin" Descriptions of Structures

The notion of spin is already well known by the students in Chemistry. The spin of the electron can be "up" or "down". In a magnetic solid, the magnetic moments of atoms can also be "up" or "down". Ferromagnetism corresponds to their parallel orientations, and antiferromagnetism to moments alternatively "up" or "down". This notion can be extended to structures based on tetrahedra, according to the direction in which the fourth vertex is oriented toward the triangular basal plane (Fig. 5.26).

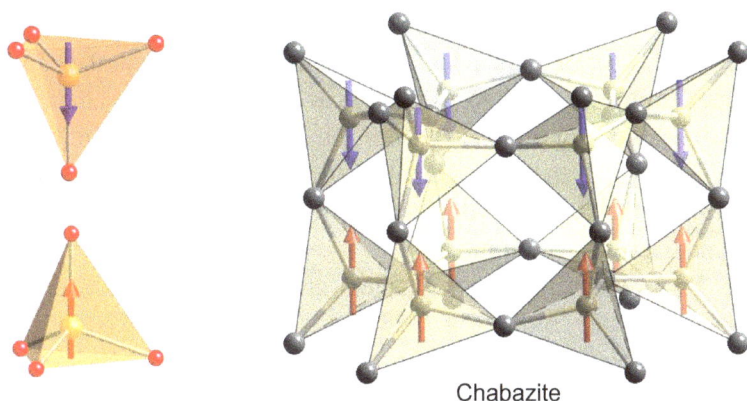

Chabazite

Fig. 5.26. (Left) Tetrahedra described in terms of spins "up" and "down" according to the direction of the fourth vertex towards the triangular basal plane; (right) illustration for the hexagonal prismatic cage (D6R) of chabazite, a hydrated sodium calcium aluminosilicate.

Despite many times hypothetical, this representation is particularly useful when simplified perspective views of structures like zeolites (which are mineral silicates and aluminosilicates) are needed. These structures are extremely complex, with, simultaneously very complicated chemical formulas. More and more, the scientists rather play on colors for differentiating the two orientations. This complexity results from extremely different orderings between the "up" and "down" tetrahedra within the same global topology. Figure 5.27 emphasizes this point by a simple example with the structures of phillipsite and gismondine.

In both structures, the basal nets of the triangles are identical and correspond to a (3.4.3.8) notation of Schläfli. The two structures differ only in the repartition

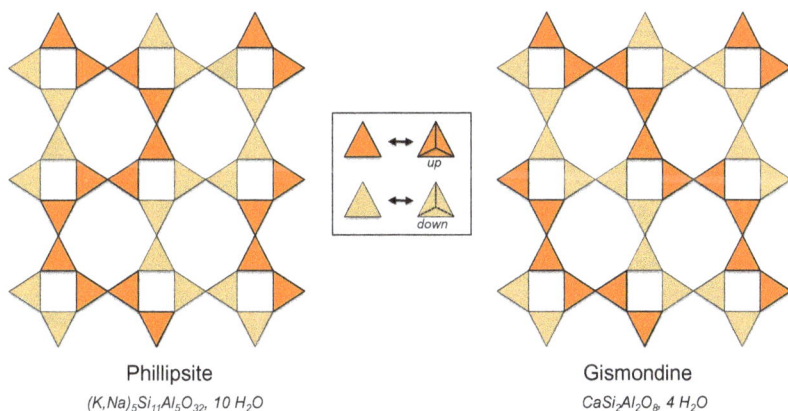

Phillipsite

$(K,Na)_5Si_{11}Al_5O_{32}, 10\ H_2O$

Gismondine

$CaSi_2Al_2O_8, 4\ H_2O$

Fig. 5.27. Idealized representation of phillipsite and gismondine in terms of "up" (orange) and "down" (yellow) tetrahedra. The chemical formulas of the two solids are indicated.

of the "up" and "down" orientations with respect to the triangular basal plane. In phillipsite, each type of tetrahedra forms sinusoïdal tetrahedral chains sharing their vertices from one chain to the other; in gismondine, the chains are quasi-linear and develop along the diagonals of the cell. Therefore, there exist two different structures within the same global topology which are only differentiated by the repartition of the "up" and "down" orientations of their tetrahedra. Otherwise, they have many points in common.

In this book, a succession of chapters allowed the reader to scrutinize the structures (for example, rutile (Fig. 5.28), under different aspects: local coordinations, various dimensionalities in the connection of polyhedra, occupation of interstices in dense packings, tiltings of polyhedra, vacancies and their influence, sliding and rotations of blocks, rod, building blocks and even spin descriptions … The reader has now at his disposal all the knowledge necessary for going farther in the relations between apparently different structures… Hence, crystal chemistry is not only the examination of structures considered as unique, on the contrary, thanks to it, Chapter 6 will establish unexpected filiations and provide unusual classifications of solids, not so easy to observe. Well understood, they will allow the chemist to come back to his primary job: the creation and transformation of new matter, but this time from rational ideas deduced from structural considerations instead of trials and errors. This will be developed in Chapter 7, the last chapter of this book. After analysis, rational creation will be on the rise…

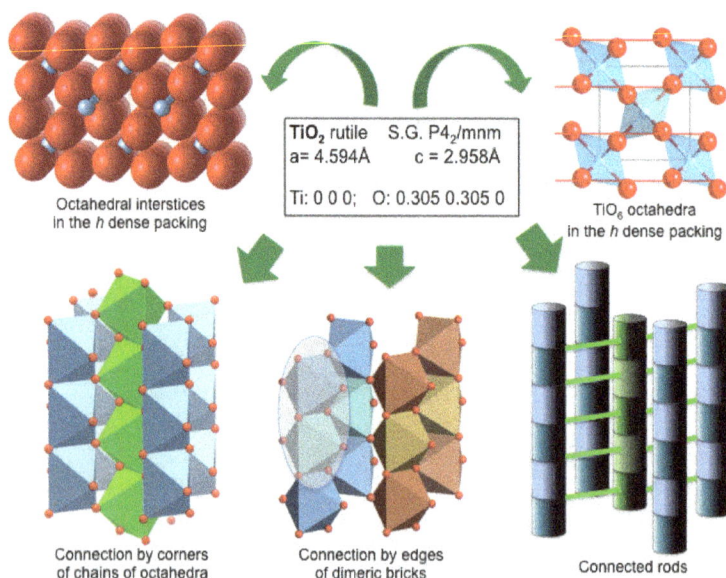

Octahedral interstices
in the *h* dense packing

TiO$_2$ rutile   S.G. P4$_2$/mnm
a= 4.594Å   c = 2.958Å

Ti: 0 0 0;   O: 0.305 0.305 0

TiO$_6$ octahedra
in the *h* dense packing

Connection by corners
of chains of octahedra

Connection by edges
of dimeric bricks

Connected rods

Fig. 5.28.   From crystallographic data to diverse complementary descriptions of the structure of titanium(IV) dioxide in its rutile form; a summary.

Chapter 6

The Structural Relations ... Their Deciphering

Chemists know that, depending on the thermodynamic conditions applied to a synthesis, an identical chemical formula can lead to different structures, which were called allotropic varieties or polymorphs of a solid (for example, carbon) in Chapter 1. These structural changes, mainly functions of temperature, time and pressure, are frequent in solid state sciences. Beyond observing them, the chemist must try to understand why these changes exist. The latter indeed result from atomic movements in the structure, which modify the arrangements in such a way that the new configuration corresponds to a lattice energy more stable than the previous one in the thermodynamic conditions fixed for the synthesis. These displacements, often complex can however be justified using the concepts already described. This chapter will try to demonstrate that, using the different modes of description detailed before, such structural transformations are easily accessible.

6.1 Structural Relations between the Polymorphs of a Solid

They can be proved by using the mechanisms of gliding or reorientation of dense packings, or by different cationic and/or vacancies orderings.

6.1.1 *Mechanisms of anionic glidings*

The most trivial example concerns the two already described polymorphs of zinc sulfide (ZnS) (Fig. 6.1). One is hexagonal (wurtzite [W]), the other cubic (blende [B]).

Only the sequences of stacking differ between the two structures: a periodicity using three different layers for blende, and two for wurtzite. The structural relation is therefore obvious: the glide of the non-superposable third layer of blende by a vector ⅓ $(a + b)$ to render it identical to the first layer.

ZnS wurtzite (hexagonal)

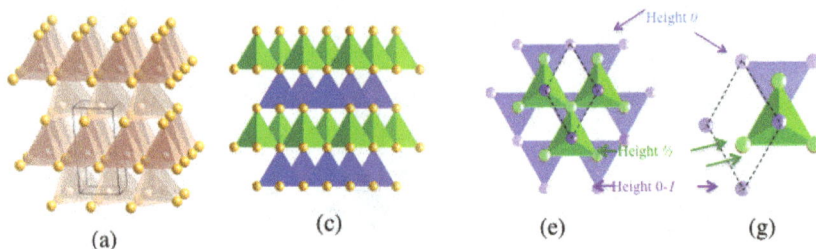

(a) (c) (e) (g)

ZnS blende (cubic)

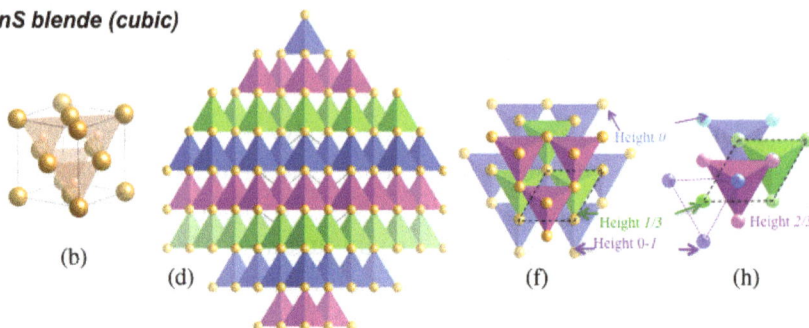

(b) (d) (f) (h)

Fig. 6.1. (a,b) Perspective views of the wurtzite (hexagonal) and blende (cubic) structures of ZnS in their polyhedral representation (corner-sharing tetrahedra); (c,d) Stacking layers of tetrahedra in both structures (two layers for wurtzite, three for blende); (e–h) succession of layers in projection.

Another example is provided by the various forms of tetravalent titanium oxide TiO_2. Besides the thorougly studied rutile form, anatase, brookite and pseudo-brookite are also natural minerals of TiO_2. Chemists, on their side, synthesized other forms, the most well known being TiO_2 B (obtained by "chimie douce" chemistry, and found later in Nature) and a high pressure form (TiO_2-II or –HP) with a structure close to that of a form of lead(IV): α-PbO_2. Its comparison with the rutile structure deserves attention.

The crystallographic data concerning TiO_2–HP are the following:

TiO_2 high pressure form		Space group: P *bcn* (**n° 60**)		
Orthorhombic	$a = 4.563$ Å	$b = 5.469$ Å	$c = 4.911$ Å	$Z = 4$
Ti(4c)	$\pm$ (0, y, ¼; ½, ½ + y, ¼)	with $y = 0.171$		
O(8d)	$\pm$ (x, y, z; ½ – x, ½ – y, ½ + z;	$x = 0.286$, $y = 0.376$,		
	½ + x, ½ – y –z; –x, y, ½–z)	$z = 0.412$		

The classical crystal chemistry study leads to $Z = 4$, $V = 122.55$ Å^3. Its cell volume is twice that of rutile (62.433 Å^3), with four motifs per cell. It exhibits four Ti-O distances at 1.91 Å, and two at 2.05 Å (with the same mean distance $\langle 1.96$ Å$\rangle$

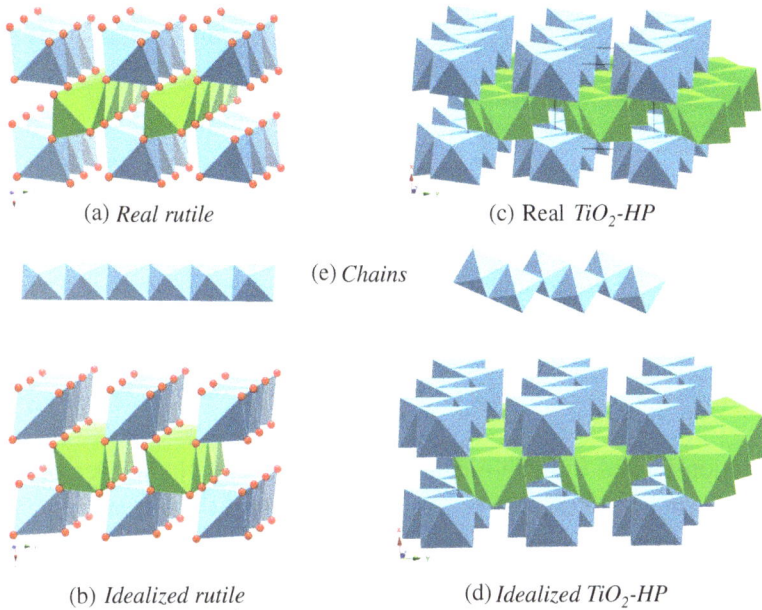

(a) *Real rutile*

(c) Real *TiO₂-HP*

(e) *Chains*

(b) *Idealized rutile*

(d) *Idealized TiO₂-HP*

Fig. 6.2.   Comparison between the TiO_2 rutile and –HP structures, real (a,c) and idealized (b,d). The central part (e) shows the different organization of the chains in both structures.

in both cases). Incidently, this shows that, at a constant formulation, pressure decreases the cell volume and increases the distortion of octahedra.

Coming back to the comparison between the two structures, their perspective views (Fig. 6.2), exhibit some similarities (same succession of quasi-compact hexagonal packings, same octahedral coordination for titanium), but also numerous differences, in particular, the organization of the chains of edge-sharing octahedra: linear in rutile, in zig-zag in HP.

How can such a transformation occur? In solid–solid transformations, except for the techniques of transmission electron microscopy, experimental techniques are not able to observe in real time the atomic movements explaining the passage from one structure to the other. The crystal chemist is therefore obliged to suggest some hypotheses for proposing reasonable movements (which mean weak magnitude) in trying to justify this passage. It is possible only if the chemist already has a good understanding of the structure.

For that, he can use an heterodox view of the octahedron. It was seen in the past that, when projected on one of its triangular faces, the octahedron appears as an hexagon. In reality, this corresponds to the superposition of two equilateral triangles of atoms rotated by 60° from one triangle to the other. The metallic (or cationic) species reside in the central interstice of the octahedron [Fig. 6.3(a)]. In other words, this means that, in layers of octahedra, each of them can be assimilated to

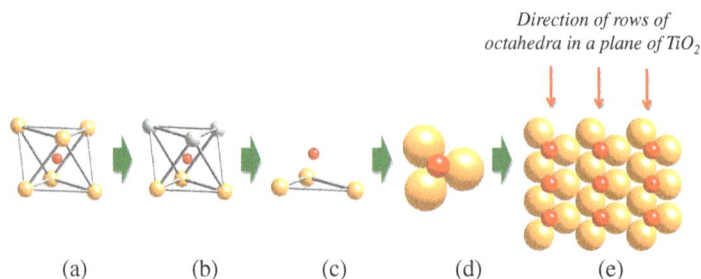

Direction of rows of octahedra in a plane of TiO$_2$

(a)   (b)   (c)   (d)   (e)

Fig. 6.3. (a) Octahedron in perspective; (b, c) its triangular reduction (considering that the grey spheres on the top belong to another layer; (d) projection of the association-related layer-metal; (e) extension giving the dense layer.

its association-related triangle-metal [Figs. 6.3(b)–6.3(d)]. A layer of octahedra [Fig. 6.3(e)] can be represented by the juxtaposition of these triangles.

   This type of view was used by B. Hyde and S. Andersson (Hyde and Andersson, 1989) for proposing an elegant mechanism for the passage rutile → TiO$_2$–HP. As in crystallographic shears, progressive anionic glides occur (Fig. 6.4). They imply the intermediary apparition of a fluorite structure, currently not isolated.

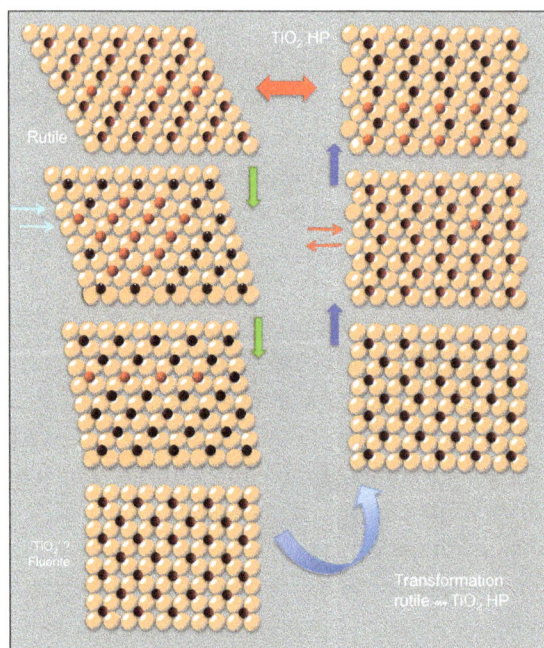

Fig. 6.4. Proposition of the progressive passage from rutile to TiO$_2$–HP by glides indicated by green and blue thick arrows. In the two figures at the top, black lines indicate the direction of the octahedral chains in the two structures. The pale blue and red thin arrows give the directions of the cooperative glides of the rows of oxide ions, in the same way as for the passage rutile–fluorite, and in the opposite direction as for the fluorite–HP one.

In the first step (rutile–fluorite), the horizontal rows of oxide ions progressively glide in the same direction (pale blue arrows in Fig. 6.4), transforming each initial triangle into a square. In the second (fluorite–HP), the displacements alternate in inverse direction (red arrows). Induced by increasing pressures, these displacements have a weak magnitude and, correlatively, are credible.

6.1.2 *Mechanisms involving rotation of blocks*

The diverse forms of manganese(IV) dioxide: MnO_2 illustrate this point. Besides pyrolusite (rutile structure), four other natural forms are known: ramsdellite (R), hollandite (H), romanechite (K) and todorokite (T).

They correspond, in a funny way, to a progressive incrementation of multichains of octahedra sharing edges … A nano Lego™ (Fig. 6.5)! For Ramsdellite, double chains connected by single rutile-like chains; in Hollandite, double chains in two orthogonal directions; the rare Romanechite associates triple chains to orthogonal double chains, triple chains which are found in the two directions for Todorokite. In this series, the associations alternatively create tunnels whose

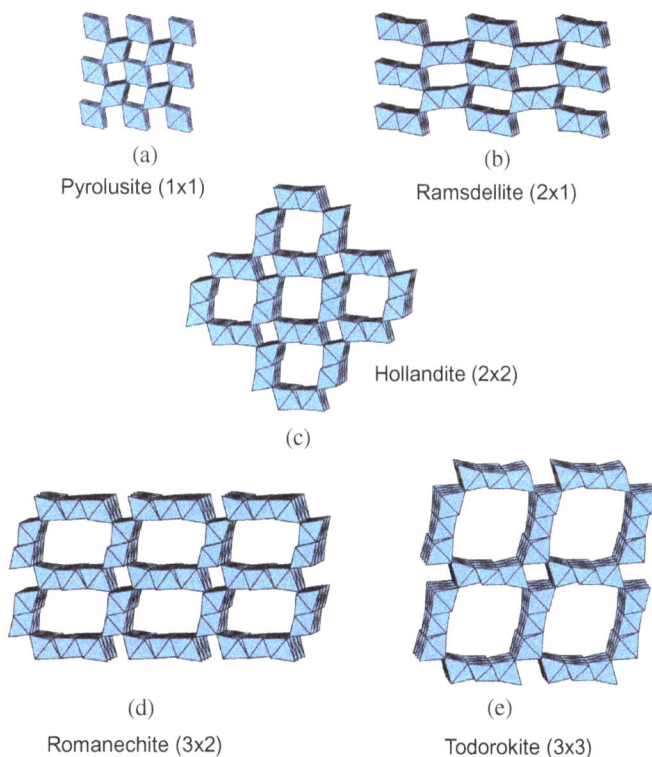

(a)
Pyrolusite (1x1)

(b)
Ramsdellite (2x1)

(c)
Hollandite (2x2)

(d)
Romanechite (3x2)

(e)
Todorokite (3x3)

Fig. 6.5.   Five polymorphs of manganese(IV) dioxide MnO_2.

sections are approximatively rectangular or square. They are usually labeled by two numbers corresponding to the number of octahedra forming the tunnel in each direction. They are noted 1×1 (rutile), 2×1 (ramsdellite), 2×2 (hollandite), 3×2 (romanechite) and 3×3 for todorokite.

The tunnels of these five forms can be completely empty, but natural minerals often host diverse cations, alkaline and earth-alkaline which modify the primitive valency IV of manganese and chemically transform this dioxide MnO_2 into mixed-valence Mn^{3+}–Mn^{4+} compounds, the Mn^3/Mn^{4+} ratio being variable, function of the amount, of the nature and of the charge of the inserted cations.

When the empty forms were considered, Sten Andersson proposed a simple and elegant relation between some of them. However, it must be noted that it is just a geometric relation which does not imply any real mechanism of solid–solid phase transition, in general, observable (if the kinetics of the transformation is appropriate) only by high resolution electron microscopy, this technique which, by Fourier calculations, allows to "see" the atoms.

The relation of Andersson implies an alternated right–left 45° rotation of rutile blocks for passing from rutile to hollandite (Fig. 6.6)... Mechanically very simple, but... unfortunately almost impossible in reality because it would imply too large displacements of the atoms in the structure! Almost impossible, but very elegant...

This concept has however an important advantage: a sort of "memory effect" in the edification of polymorphs, a common denominator for apparently different structures.

It is also valid for a MX_3 network with the transformation of perovskite into TTB, the tetragonal bronze of Magnéli.

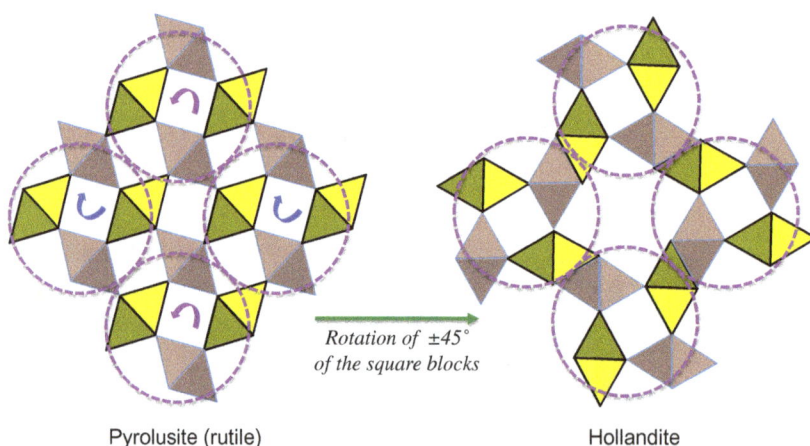

Rotation of ±45°
of the square blocks

Pyrolusite (rutile)   Hollandite

Fig. 6.6. Possibility of passage from rutile to hollandite by the clockwise and counterclockwise alternated rotations (indicated by purple and blue arrows) of the square blocks inside the dotted circles. In the blocks, yellow octahedra are at a height ½ and those in orange at heights 0 and 1.

A$_{0.6}$MX$_3$ perovskite   K$_{0.6}$MX$_3$ TTB

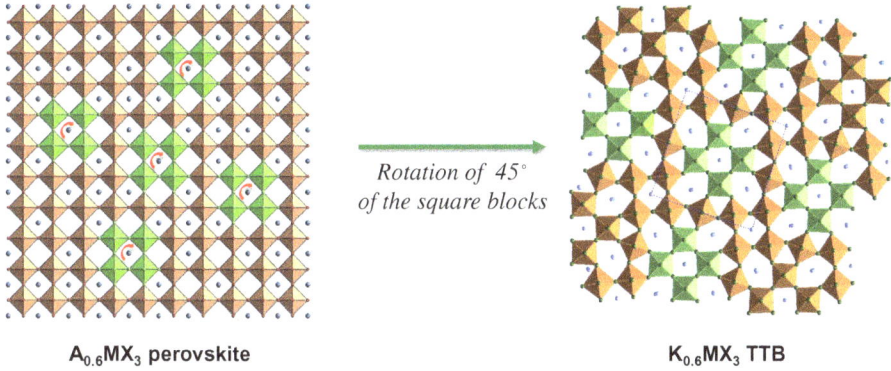

Fig. 6.7.   Defective perovskite to TTB transformation by rotation of 45° of green blocks.

It is worthy to note that, with these rotations, two symmetric operations are possible [Fig. 6.8(a)]. Starting from the same ReO$_3$ matrix, but by choosing the type of block submitted to the roration, two different ways lead to the TTB topology. This implies a possibility of nano-twinning at the beginning of the growth of crystals [Fig. 6.8(b)], either by coexistence of two particles or interpenetration of two domains with the same composition and structure. Twinning is one of the structural mechanisms for decreasing the stress during the growth.

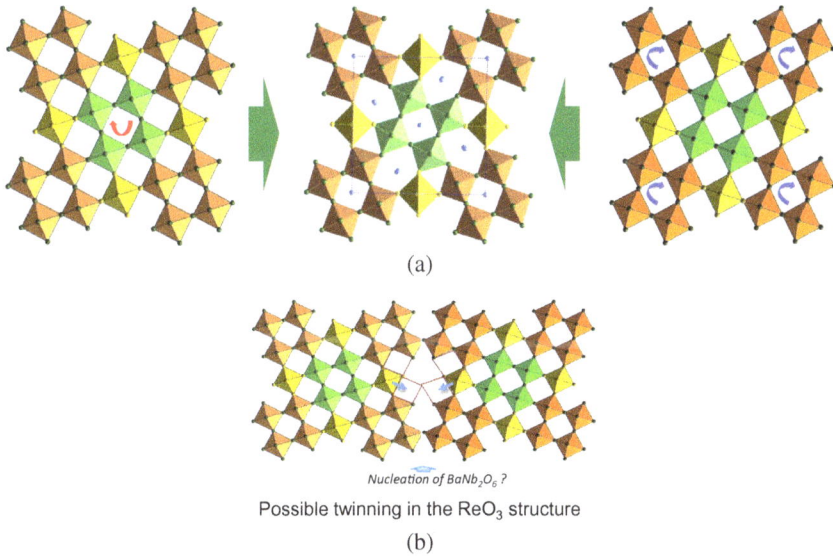

(a)

Nucleation of BaNb$_2$O$_6$?

Possible twinning in the ReO$_3$ structure

(b)

Fig. 6.8.   (a) The two ways of access to TTB by the rotation of the blocks 2 × 2 of the ReO$_3$ structure; the yellow octahedra participate in the rotation; (b) representation of the possible twinning which could be the initiation of a BaNb$_2$O$_6$-type nucleation after the local glides of the dotted octahedra.

The space between the two parts, as described in Fig. 6.8(b), is large enough for the insertion of octahedra, resulting from a local glide between the two parts of the twin. In this case, the local structure could be reminiscent of the $BaNb_2O_6$ structure (see Nb_2O_6 framework in Chapter 5), with pentagonal tunnels; this twinning could induce the nucleation of this phase. This unlikely situation gives the opportunity of a wise remark… Never allow crystal chemistry to declare more than it can do! It remains for the moment a description which must always be related to a reasonable chemistry. It is indeed more likely that, if there is a "memory effect", it would rather exist in the R form of Nb_2O_5 (Paving 4.1) which already exhibits edge-shared dimers.

6.2 Structural Relations between Solids with Different Formula

Different formulas, which essentially take into account the chemical nature of the atoms in the solid, do not always mean different topologies. This was already proven during the study of di- and trirutiles (Fig. 4.14) which all have the rutile topology, but with at least one cell parameter multiple of the pure rutile one. The **geometric ordering** between the occupied sites of the structure modifies the cell parameters, sometimes the cell symmetry, and induces some anionic displacements, at invariant topology. The different cations can be identical or present different valencies. The type of ordering then depends first on the ratios between the cations in the chemical formula.

Two series of examples, coming from already known structures, will illustrate this part: the perovskite (Mitchell, 2002) and the spinel structures.

6.2.1 *Cationic ordering in perovskites*

One already knows the great chemical adaptability of this structure, able to host many kinds of cations with different nature and charges. It accepts significant deviations to the structural ideality, through the tolerance factor of Goldschmidt, related to the respective sizes of the cations involved in the structure. The larger their contrast, the larger the deformation at constant topology. This introduces diverse tiltings mainly influencing the positions of anions and the coordination of A ions, although ensuring always a three-dimensional association of corner-shared octahedra. This cationic ordering will have consequences in terms of symmetry. The cell can remain cubic, but can also change, up to the triclinic symmetry. This symmetry will be a function of several parameters: chemical and geometrical nature of the ordering, the localization of atoms of the same type on the concerned sites and, of course, the type of tilting, among the 23 possibilities of Glazer.

Ordering has therefore a major effect.

Chemical formulas as different as $BaBiO_3$, Sr_2DyRuO_6, Ba_2NaReO_6, Na_3AlF_6, $Ba_3Al_2WO_9$, $Sr_3CoSb_2O_9$, $Ca_4Nb_2O_9$, $NaMn_7O_{12}$, $CaCu_3Ti_4O_{12}$, $Na_3La_3Mg_2Ta_4O_{18}$ cover the same reality: they all belong to the AMX_3 perovskite topology, with its two types of sites: A sites, with their initial twelve coordination, and M sites with an octahedral environment, regular or not. The creation of a cationic ordering in it requires at least two types of cations in either one or both sites, and also, simple ratios between them.

But what kind of ordering? If cations are ordered on only one site (A or M), we will then have **double perovskites**. The reader must not apply this term in a too strict sense. Indeed, the cations or anions ordering in the perovskite structure is described, **depending on the authors** through the order (their alternation along 1, 2 or 3 directions) denoted as 1.1, 1.2... and/or the multiplicity of the resulting cell parameters. As an example, in $YBa_2Cu_3O_7$: only one variety of M atom, two sites for M, and triple c parameter (cf. page 171 and others). They are said to be complex (or quadruples) if ordering occurs simultaneously on both sites.

If order occurs exclusively on the octahedral M site, the most frequently encountered formulas will be $A_2MM'X_6$ (or $AM_{0.5}M'_{0.5}X_3$ by reference to the AMX_3 general formula), or $A_3MM'_2X_9$ (or $AM_{1/3}M'_{2/3}X_3$). These two types of order on M are labelled 1:1 and 1:2, respectively. Note that for the last, the order, apparently 1:2, can, in reality, be a 1:1 order if one of the cations is distributed on both M and M' sites (e.g. $Ba_3UFe_2O_9$). The developed structural formula taking into account the real order is $([Ba_2][Fe_{1/3}U_{2/3}][Fe]O_6)$ since iron exists on both M and M' sites.

A 1:1 ordering can also be observed on A sites (formula $AA'M_2X_6$), and a 1:3 order also often exists $[(AA'_3)(M_4)O_{12})]$.

6.2.1.1. *Double perovskites with a 1:1 order on M site*

According to the charge of A, the corresponding solids can be classified into two groups:

- If A is divalent (Ca, Sr, Ba), four associations are possible, depending on the charges of M and M':

  $A^{2+}_2M^{3+}M'^{5+}O_6$; $A^{2+}_2M^{4+}M'^{4+}O_6$; $A^{2+}_2M^{2+}M'^{6+}O_6$; $A^{2+}_2M^{1+}M'^{7+}O_6$;

- If A is trivalent, three theoretical possibilities:

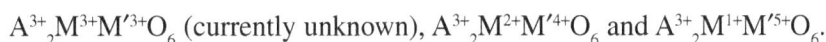

  $A^{3+}_2M^{3+}M'^{3+}O_6$ (currently unknown), $A^{3+}_2M^{2+}M'^{4+}O_6$ and $A^{3+}_2M^{1+}M'^{5+}O_6$.

Two types of ordered organization then exist (Fig. 6.9). In terms of symmetry, Type I is the richer. Depending on the constituents, all the crystalline systems

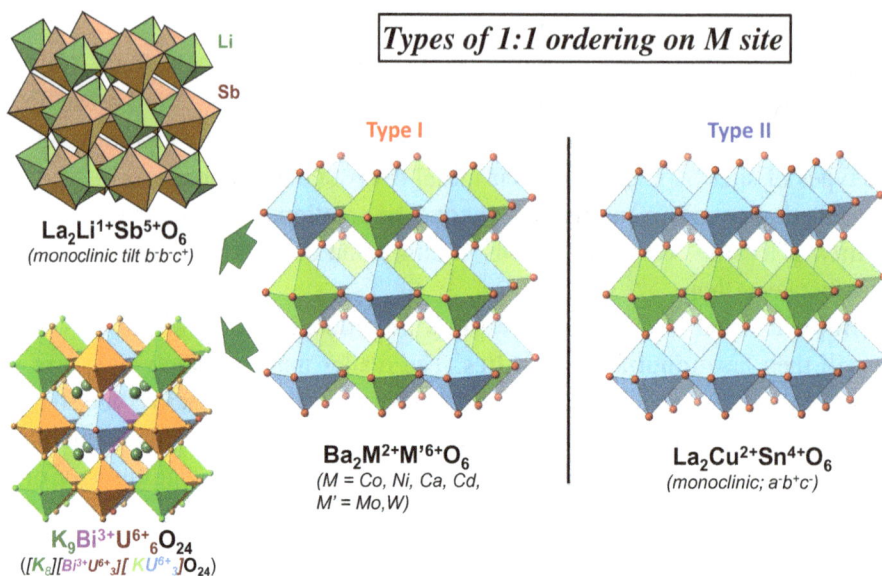

Fig. 6.9. Illustration of two types of 1:1 order on M sites. In the center, $Ba_2MM'O_6$ (order by centered faces) and, on the right, the ideal La_2CuSnO_6 (layers ordering) where tilting is not taken into account. Both figures on the left present the same type of fcc order in M and M'. La_2LiSbO_6 is monoclinic and tilted; the other, $K_9BiU_6O_{24}$ is cubic. In it, a secondary order exists, within the sites M and M' (see developed formula).

are represented. It is the mode of tilting influenced by the A cation which is the pertinent parameter.

Concerning this point, the tolerance factor of Goldschmidt was not considered as sufficiently discriminating for taking into account the evolutions of symmetry in double perovskites. A new factor, proposed by Teraoka and labeled the *fitness factor* Ω (Mitchell, 2002). The expression of this factor is $\Omega = R_A \sqrt{2} / (R_M + R_X)$. In it, R_M represents the average ionic radius on sites M and M'. In several families of double perovskites, this explains well the progressive evolution from a cubic symmetry to a monoclinic one, through tetragonal and orthorhombic systems. The cubic symmetry corresponds to the largest values of Ω ($1.13 \geq \Omega \geq 1.03$); the solids adopt the tetragonal system for $1.02 \geq \Omega \geq 0.97$; they are orthorhombic for $0.96 \geq \Omega \geq 0.92$. Below, the symmetry is monoclinic. Figure 6.9 provides two significant examples (except triclinic).

The order in $K_9BiU_6O_{24}$ is rare and deserves attention. Within the global 1:1 order, a secondary order appears on sites M and M'. Uranium is equally distributed between M and M', with a 1:3 Bi/U sub-order in M site and a similar one K/U in site M'. This means that K^+ ions occupy two notably different types sites (as in cryolith K_3AlF_6): completely filled dodecahedral 12 coordination on A site and partially on M' sites.

Type II is more rare, and is often observed when large trivalent ions occupy A sites.

6.2.1.2 *Double perovskites with a 1:2 order on site M*

First discovered by Galasso in 1959, their general formula is $A^{2+}_3M^{2+}M'^{5+}_2O_9$. At this point, the reader must be careful about an eventual reciprocity between the formula and the order: indeed, even if a solid has the formula $A_3MM'_2X_9$, it does not necessarily mean that it exhibits a 1:2 order. This can also correspond to a 1:1 order. For example, in $Ba_3WAl_2O_9$, ¼ of the Al^{3+} ions are statistically distributed with W^{6+} ions on the M site and completely occupy the site M'. The developed formula $[Ba_3][(Al_{1/3}W_{2/3})]_{1.5}[Al]_{1.5}]O_9$ indicates a 1:1 order. The same case occurs with $Sr_3CoSb_2O_9$, antimony alone occupying the site M'.

When a real 1:2 order exists in a perovskite, the cell is rhomboedral and described in its multiple hexagonal cell. Figure 6.10 shows the oxide $Ba_3SrTa_2O_9$ represented in the multiple hexagonal cell of the rhombohedron. The large Ba is on the A site and the smaller Sr in octahedral coordination. The 1:2 order corresponds to the alternation of single octahedral sheets of strontium and of double layers of tantalum octahedra. In such a structural type, tiltings of octahedra can occur when small divalent cations (e.g. Mg^{2+}) occupy the M site.

6.2.1.3 *Double perovskites with a 1:3 order on site M*

This case is very rare [Fig. 6.10 (right)] with only two known examples when the small alkaline ions Li^+ and Na^+ are involved. These cations are localized only at the vertices and the center of the cell. The solids therefore exhibit a *I* lattice. In such order, the presence of very polarizing Sb^{5+} probably explains this type of ordering. Indeed, a substitution of Sb ions by niobium and tantalum ones lead to

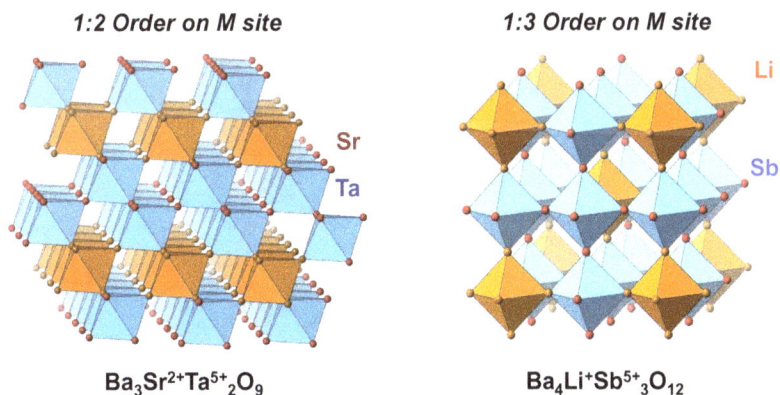

1:2 Order on M site   **1:3 Order on M site**

$Ba_3Sr^{2+}Ta^{5+}_2O_9$ $Ba_4Li^+Sb^{5+}_3O_{12}$

Fig. 6.10.   Illustration of the 1:2 (left) and 1:3 (right) orders on M site.

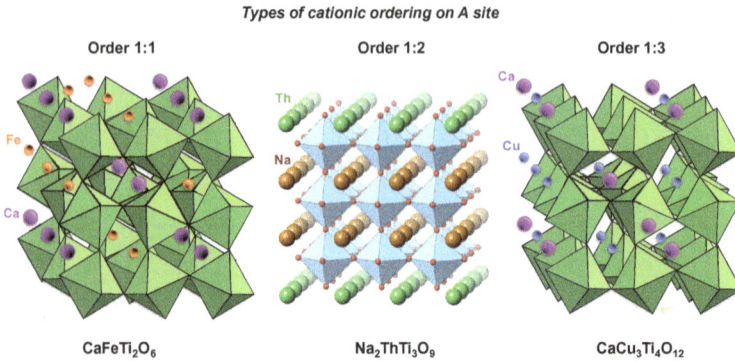

Types of cationic ordering on A site

Order 1:1   Order 1:2   Order 1:3

CaFeTi$_2$O$_6$ Na$_2$ThTi$_3$O$_9$ CaCu$_3$Ti$_4$O$_{12}$

Fig. 6.11.   Illustration of the 1:1, 1:2 and 1:3 orders on A site.

disordered phases. Note that the replacement of barium by strontium leads to a monoclinic symmetry.

The above examples are exclusively concerned with the M subnetwork. What is the situation when ordering occurs on the A sites?

6.2.1.4 *Perovskites and cationic ordering on the A sites*

Here also, the 1:1, 1:2 and 1:3 orders coexist, but examples are extremely rare (Fig. 6.11).

6.2.1.5 *Ordered complex perovskites (orders on both A and M sites)*

Only one example is clearly identified: (NaLa)(MgTe)O$_6$ which presents a 1:1 order on both A and M subnetworks (Fig. 6.12).

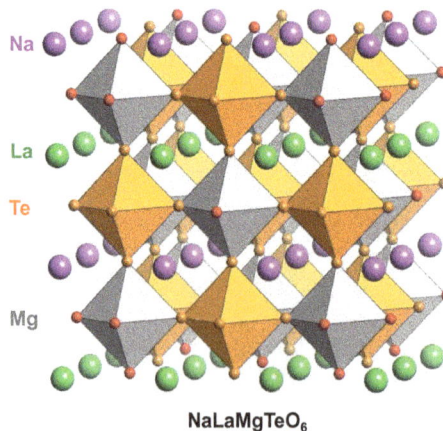

NaLaMgTeO$_6$

Fig. 6.12.   1:1 order existing simultaneously on sites A and M. It also exists when potassium replaces sodium on the A site.

On the M site, the 1:1 order gives a repartition identical to that existing for ions Na^+ and Cl^- in NaCl. On the contrary, sodium and lanthanum ions on A sites are both located in alternate layers. This ordered distribution is probably due to the large differences between the charges of monovalent sodium and trivalent lanthanum.

6.2.2 *Cationic ordering in spinels MM'_2O_4*

This complex structure was studied in detail in Sec. 3.2.5. Whatever its type of description (crossed rods, stacking of layers), the common point is the existence of two types of sites: octahedral and tetrahedral in a 2:1 ratio. Two limit cases of occupation occur for the corresponding trimer: a full occupation by one type of ion of each site; the spinel is found to be direct. The inverse spinel corresponds to the other case (Fig. 6.13).

They are effectively limiting cases. With time, all the cases of distributed occupations between these two extremes have been encountered experimentally by chemists and physicists. After Néel, the latter were particularly interested by their fantastic magnetic properties. These spinel oxides belong to the best materials ever discovered since the ancient Greek civilization and the discovery of the mineral magnetite Fe_3O_4, which behaves as a magnet up to 600°C. The variety of magnetic couplings allowed by the huge adaptability in compositions and in the distributions on the two sites has allowed an almost continuous tuning of the magnetic properties up to 700°C. It is during his study of this family of oxides that

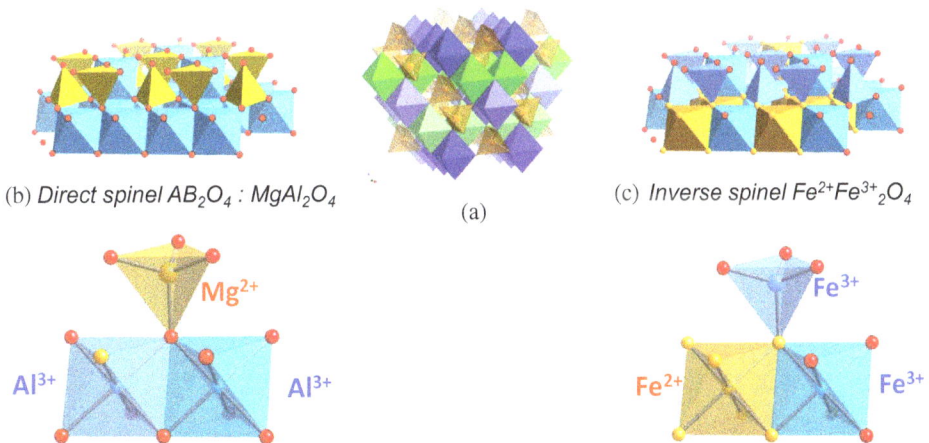

(b) *Direct spinel AB_2O_4 : $MgAl_2O_4$*   (a)   (c) *Inverse spinel $Fe^{2+}Fe^{3+}_2O_4$*

Fig. 6.13.   The two types of occupation of the two cationic sites in direct (left) and inverse spinels (right). In (a) is recalled the rod description of spinel with different codes of color for each height.

Louis Néel dicovered and explained a new magnetic behavior: ferrimagnetism. For that, he received the Nobel Prize in Physics in 1971.

From the above paragraphs, it is now clear that cationic ordering, even in well known topologies, has an incidence on the chemical compositions, on the evolution of structures (due to tiltings and cation sizes), and also on the physical properties of the considered solids.

6.2.3 *Ordering between ions and vacancies*

This aspect was already developed when speaking about non-stoichiometry in Chapter 4, and about brownmillerite-like structures. It is therefore not necessary to multiply the examples in this initiation, except for one case which concerns copper oxides CuO_x (Fig. 6.14) because it shows that crystal chemistry allows the access to mechanisms of formation during gas–solid reactions.

It is a good example of a **topotactic** transformation. In this series, the organization of copper ions is not affected during the reaction. Only the oxide ions arrangement changes as a function of the increasing amount of oxygen atoms, above and in the solid. At the beginning of the oxidation of copper ($Cu^0 \rightarrow Cu^+$), cuprite Cu_2O is formed; copper becomes a monovalent cation, which is linked to two opposite oxide ions (dumbbell coordination), and the three-dimensional

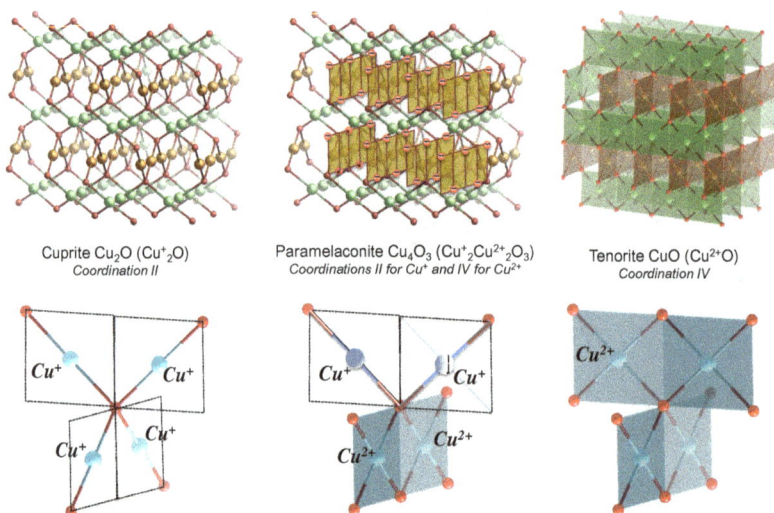

Cuprite Cu_2O (Cu^+_2O)
Coordination II

Paramelaconite Cu_4O_3 ($Cu^+_2Cu^{2+}_2O_3$)
Coordinations II for Cu^+ and IV for Cu^{2+}

Tenorite CuO ($Cu^{2+}O$)
Coordination IV

Fig. 6.14. Evolution of the coordination of copper during the progressive passage from the structure of cuprite Cu_2O (monovalent copper) to paramelaconite Cu_4O_3 (mixed-valence Cu(I)/Cu(II)) and finally to tenorite CuO containing only copper(II).

framework results from the orthogonal cross-linkage in two directions of the space of infinite zig-zag chains O–Cu–O ... The oxygen atoms are tetrahedrally coordinated, at the crossing between two orthogonal chains.

When the oxidation process continues, one-half of the cuprous ions Cu^+ is transformed into cupric ones Cu^{2+} and form paramelaconite: Cu_4O_3, or $(Cu^+)_2(Cu^{2+})_2O_3$, a rare mixed-valence mineral. In its structure, the zig-zag chains in one direction are transformed into infinite chains of edge-shared square planes occupied by Cu^{2+}, while the other one, perpendicular, is always built from zig-zag chains like in Cu_2O. The change of coordination is complete when oxidation is completed with the formation of tenorite CuO. All the chains, primitively in zig-zag, become infinite linear chains of edge-shared square planes occupied by Cu^{2+}, without any modification of the copper subnetwork. Only the coordination changes.

This being observed, a tentative mechanism for the successive oxidations of metallic copper can be proposed. Careful experiences on the oxidation of copper showed that it is progressive, layer by layer, with, on the surface, the successive appearances of cuprite layers Cu(I), then paramelaconite (Cu(I)/Cu(II)) and finally tenorite Cu(II). This implies reactions at the surface, before a diffusion of oxide ions in the metallic matrix. It must be understood that, during the first steps of the action of oxygen at the surface of the face centered cubic copper, subtle structural perturbations occur at the ångström scale (Fig. 6.15).

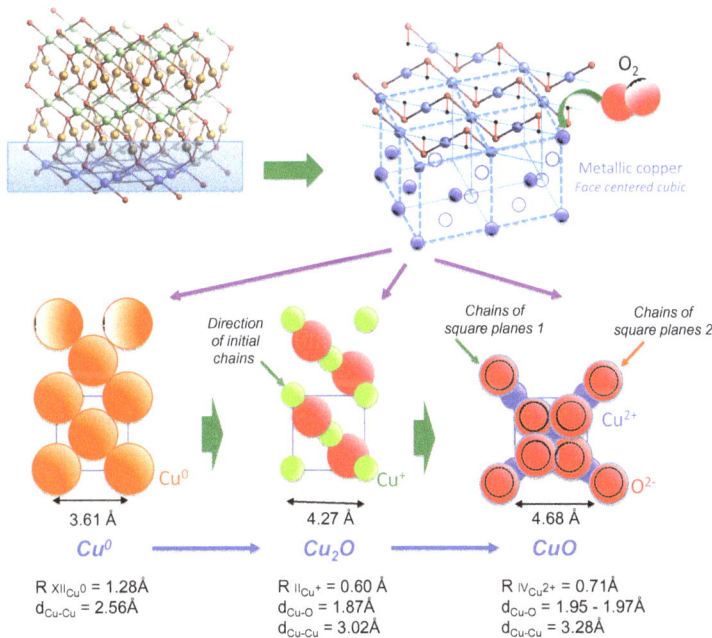

Fig. 6.15.   Structural evolutions at the surface of copper during the first steps of its oxidation.

The gaseous dioxygen molecules which arrive on the surface are transformed into oxide ions ($R_{O^{2-}}$ = 1.21 Å). Copper, which was in 12-fold coordination in the metal (metallic radius 1.28 Å) becomes copper(I) in two-fold coordination II with a smaller ionic radius (0.60 Å). The insertion of oxide ions induces a dilatation of more than 18% of the cell parameter. A look at the Cu_2O structure provides the location of these inserted anions O^{2-} between two cations Cu(I) along *only one* face diagonal of the *fcc* structure of copper. They form the first zig-zag chains in only one direction. Further, the arrival of other O_2 molecules will form another layer with perpendicular chains. It is when the crossed zig-zag chains are formed on a certain thickness that supplementary oxygen atoms will enter and oxidize one-half of the cuprous ions and transform the two-fold coordination of copper(I) into a square plane by the insertion of new oxide ions, first in one direction, forming paramelaconite. The end of the oxidation leads to tenorite CuO, with all the Cu^{2+} ions in a square planar coordination IV.

This remark shows how a very careful crystal chemistry examination allows to overcome the purely analytic description for proposing, in a comparative dynamic approach, a structural image of some mechanisms of reaction (here oxidation).

The structural approach of the oxidation of copper is not a unique example. Its application to the oxidation of iron allows to go further. At room temperature, this metal exists in its α form, centered cubic. This means that the attack on the faces by oxygen will be different (Fig. 6.16). These faces represent a square of non-tangent atoms, which will be the nucleus of the formation of oxide ferrous "FeO" *fcc* (the stoichiometric chemical formulation used here is not correct). This solid is indeed never strictly stoichiometric and would better be written as $Fe_{1-x}O$. The inverted commas are used to signify such purpose. However, for the sake of simplicity, this simplified notation will be retained in the following.

The first layer of oxide ions initiates the germination and growth of "FeO" on Feα. The ferrous oxide exhibits a *fcc* structure of the NaCl-type. The Fe^{2+} ions are at the center of edge-shared octahedra [Fig. 6.16(b)], and ferrous oxide will be the first step of the oxidation of iron. The total oxidation will correspond to the spinel γ-Fe_2O_3, with exclusively Fe^{3+}, after the intermediary apparition of the mixed-valence oxide Fe_3O_4, the famous magnetite. Therefore, a question: can crystal chemistry help in a simple molecular understanding of what is observed at the macroscopic scale?

It is noted, that at the beginning of the oxygen attack, one oxygen ion arrives at the center of the square of non-tangent Fe^0 and induces a 1:1 Fe/O ratio [Fig. 6.16(a)], which nucleates the growth of "FeO". A first look at the developed formula of iron oxides helps. The sequence of oxidation:

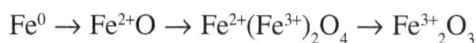

$$Fe^0 \rightarrow Fe^{2+}O \rightarrow Fe^{2+}(Fe^{3+})_2O_4 \rightarrow Fe^{3+}_2O_3$$

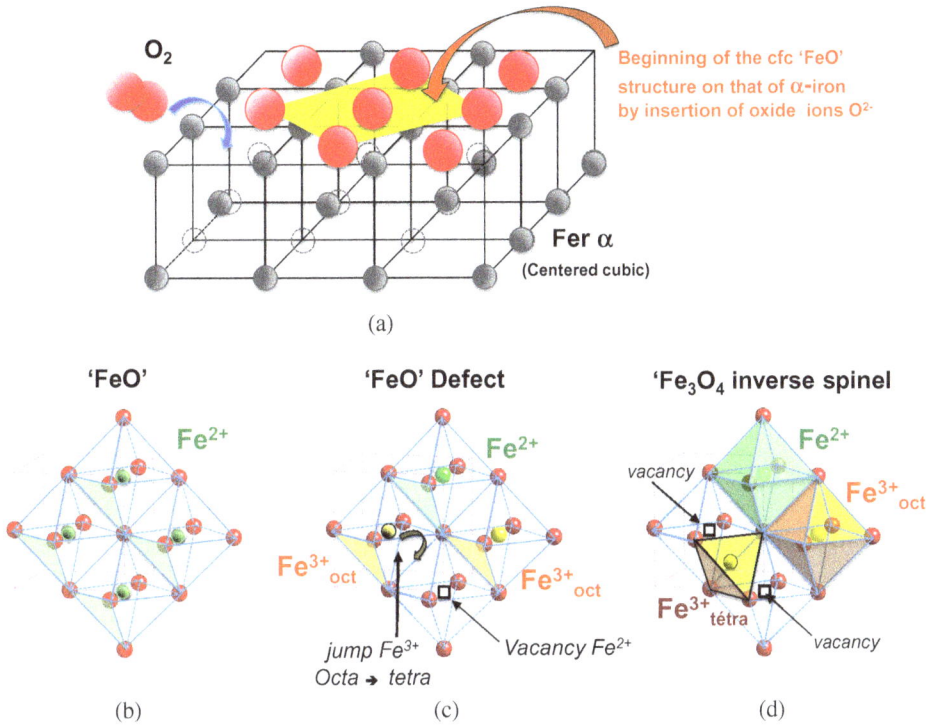

Fig. 6.16. (a) Illustration of the localizations of the species during the attack by oxygen on the surface of iron α. The oxide ions, when arriving at the center of the square of non-tangent iron atoms of the surface, adopt a *fcc* arrangement which serves to the nucleation of "FeO" (NaCl type); (b) perspective view of a fraction of the "FeO" cell with two edge-sharing Fe^{2+} octahedra; (c) the creation of a vacancy in the "FeO" framework generates, with respect to electroneutrality rules, the transformation of two Fe^{2+} neighboring the vacancy into two Fe^{3+} octahedra (d) followed by the jump of one Fe^{3+} from an octahedral site to an intersticial tetrahedral one. The iron trimer, characteristic of the building block of an inverse spinel (Fig. 6.13), is obtained.

can be written at constant amount of oxygen in the formula:

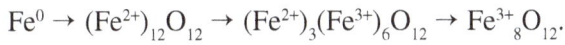

$$Fe^0 \rightarrow (Fe^{2+})_{12}O_{12} \rightarrow (Fe^{2+})_3(Fe^{3+})_6O_{12} \rightarrow Fe^{3+}_8O_{12}.$$

And, globally,

$$Fe^0 + oxygen \rightarrow Fe_{12}O_{12} \rightarrow Fe_9O_{12} \rightarrow Fe_8O_{12}.$$

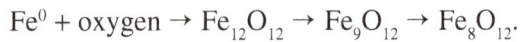

Looking at the structure of the three oxides, all face centered cubic, and the oxygen subnetwork, also always *fcc* whatever the development of the oxide layers, it must be concluded that the passage from Fe^0) to Fe_2O_3 occurs by creation and migration of cationic vacancies [Figs. 6.16(b)–6.16(d)].

If, starting from the square of four Fe^{2+} ions of Fig. 6.16(b), a vacancy is created in one of them, the obligation to respect the electroneutrality implies for two nearest neighbors Fe^{2+} to change their valency and oxidize into ferric Fe^{3+}. For reasons of minimization of the lattice energy, one of the Fe^{3+} ions will jump from its initial octahedral site to the closer tetrahedral intersticial site [Fig. 6.16(c)] and create the configuration represented in [Fig. 6.16(d)] with one ferrous Fe^{2+} and one ferric Fe^{3+} ion in octahedral sites and one Fe^{3+} in one tetrahedral site. It is exactly the repartition observed in the inverse spinels represented in Fig. 6.13. Finally, the passage from Fe_3O_4 to ferric oxide γ-Fe_2O_3, which contains exclusively Fe^{3+} ions, retains the spinel topology through the creation of a supplementary cationic vacancy.

This question on the mechanism of oxidation of iron — which differs from that of copper because the structures of the starting metals are different — have then allowed to establish a structural relation between NaCl and spinel structures. It was not mentioned up to now. The second conclusion of this paragraph globally concerns the structures of iron oxides. The percentage of vacancies increases with oxidation. Such a conclusion could not have been extracted without a careful knowledge of the three structures.

6.3 The Inverse Structures

Through this book, chemists begin to understand that the chemical formula, even if it informs on the ratios (stoichiometric or not) between the elements of the solid, does not (most of the time) shed some light on its structural description. The inverse structures provide a funny illustration on this point.

Take the nitride Cu_3N. At variance to oxides, this more and more studied nitride family was not already mentioned in this book, but the rules defined for oxides apply also to this family. Question: is Cu_3N a new structure type? Clearly not! It adopts the trivial ReO_3 topology (Fig. 6.17); N^{3-}, the nitride anion, occupies the place of rhenium, and copper(I) the place of oxide anions O^{2-}. Cationic and anionic sites are just inverted in an invariant topology. It is a supplementary proof of the strong influence of lattice energy, which governs the structure types and preserves the most usual coordinances of every element.

The second example returns to cuprite Cu_2O (p. 117) and its Cu(I) crossed zig-zag chains. A focus on oxygen instead of Cu shows it is surrounded by four Cu neighbors that form corner-shared tetrahedra $OCu_{4/2}$ like the $SiO_{4/2}$ ones in β-cristoballite, a variety of silica SiO_2. Here also, occurs the inversion of the cationic and anionic sites and invariance of the topology.

ReO₃ topology

β-cristoballite topology

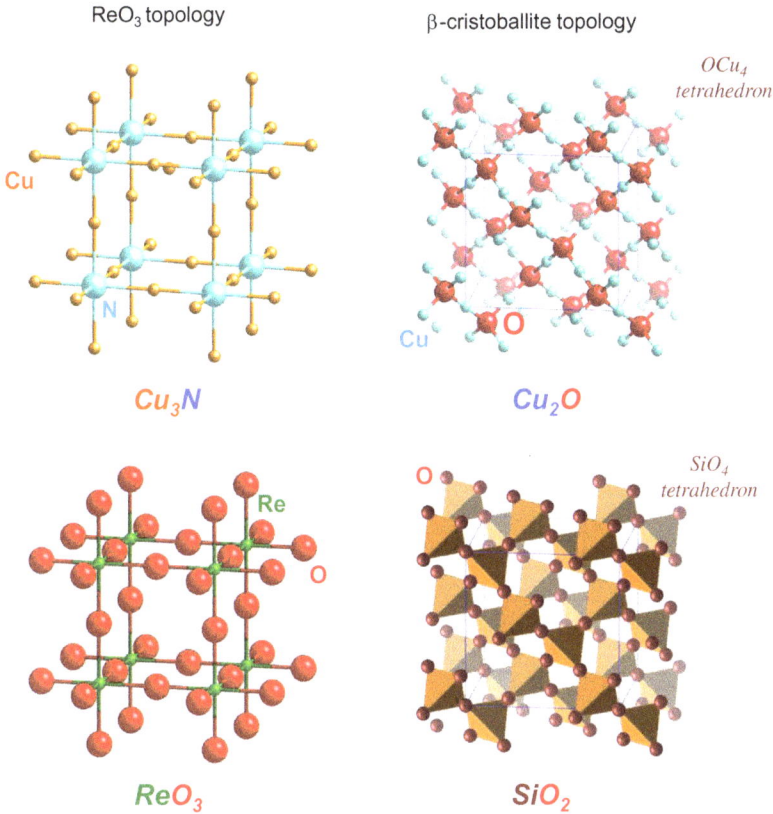

*OCu₄
tetrahedron*

Cu

N

Cu₃N

Cu

O

Cu₂O

Re

O

ReO₃

O

*SiO₄
tetrahedron*

SiO₂

Fig. 6.17.   Two examples of inverse structures: cationic and anionic sites are inverted while keeping the same topology.

From this, what can be the information extracted by the crystal chemist? He must not focus on the precise chemical formula of a solid, but rather concentrate on its general formula... If Cu_3N was initially considered as belonging to the general family MX_3, the surprise would have been less significant! ...

6.4 Interconnection of Different Building Blocks

Up to now, the brick description of simple structures was treated as the different organizations of identical bricks. However, like architects who, depending on the eras, or their fantasy and their budget, built houses (Fig. 6.18) by using different materials, Nature and chemists can do the same (Figs. 6.19 and 6.20).

Fig. 6.18.   Houses built with bricks, (left) identical or (right) different.

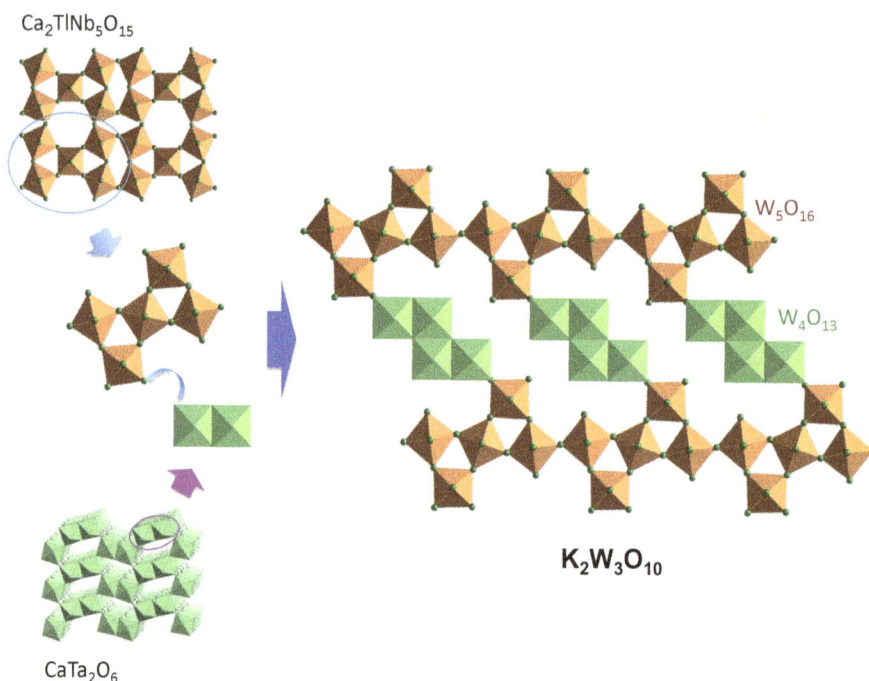

Fig. 6.19.   Construction of the framework of $K_2W_3O_{10}$ (or $K_6W_9O_{30}$) from the pentamer M_5O_{15} and the dimer M_2O_6 (already observed in $CaTa_2O_6$ and $BaNb_2O_6$). The formula of these two types of blocks are indicated, taking into account the connection of these bricks which have terminal vertices. The sum of these formula gives 29 oxygen atoms. The last one lies in the tunnels. Associated with K^+, it forms infinite K–O–K... chains which develop perpendicularly to the plane of projection.

The first example concerns $K_2W_3O_{10}$, built from heptamers (Fig. 6.19) (Rao and Raveau, 1995). The latter result from the connection between the pentamers already observed in $Ca_2TlNb_5O_{15}$ (Fig. 6.21) and the dimers (see Fig. 6.20) existing in $CaTa_2O_6$ and/or $BaNb_2O_6$ structures.

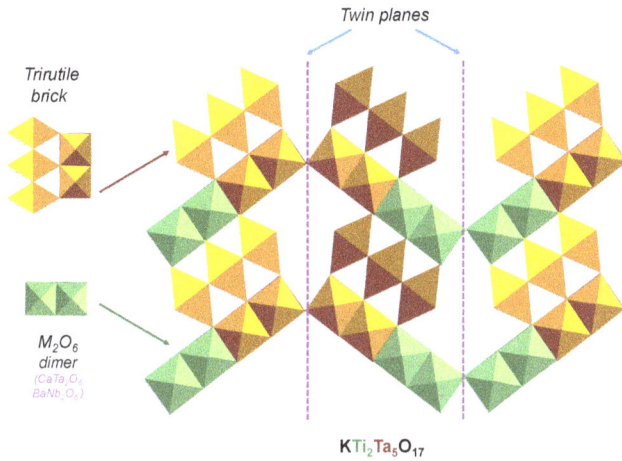

Fig. 6.20. Construction of the framework of $KTi_2Ta_5O_{17}$ from trirutile and dimeric blocks. The connections between blocks can be considered as a periodic micro-twinning between $Ti_2Ta_5O_{17}$ heptamers. The twinning plane is shown with dotted lines.

The second example, the titano-tantalate $KTi_2Ta_5O_{17}$ (Fig. 6.20), associates trirutile blocks and the above dimers. These two blocks are linked by vertices and edges (Rao and Raveau, 1995).

The third example relates once more to the pyrochlore structure (Fig. 6.21). It was already thoroughly analyzed in different ways, but the current way of description was missing.

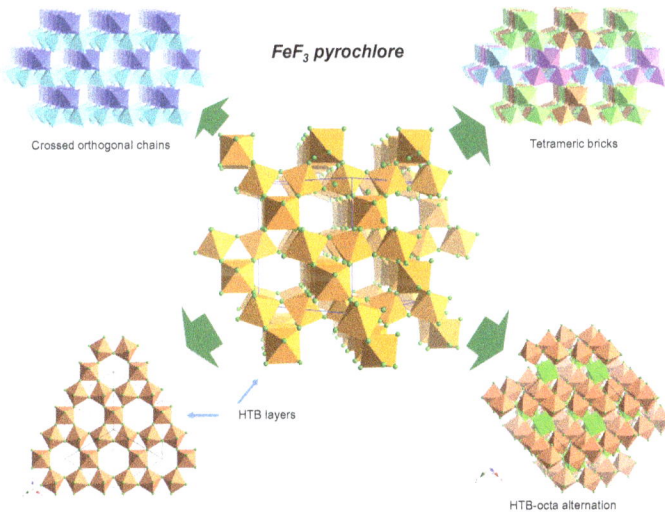

Fig. 6.21. The diverse descriptions of the pyrochlore structure. The two figures in the bottom part represents the HTB layers (left) and their corner-shared associations (right) with isolated octahedra (in green).

Indeed, one of its characteristics is the existence of layers of the HTB type, in the (111) planes of the cubic cell. Isolated octahedra connect these layers for ensuring the three-dimensional framework. One must admit, however, that this description is rather artificial when compared to that in tetrahedra of octahedra. Indeed, the current description with two types of bricks implies to give a special role to the isolated octahedra. This has no crystallographic reality because all the iron atoms are only on a single crystallographic site.

The last example: $Rb_3Nb_{54}O_{146}$ (often called the Gatehouse bronze from the name of its author) is complex but fascinating (Fig. 6.22). Indeed, its description needs the use of several sets of bricks, which were all described in this book. It is extremely rare (Rao and Raveau, 1995).

This extremely rare structure provides, like hollandite, an example of tunnels formed by more than six octahedra. Note that the distortion of some octahedra is

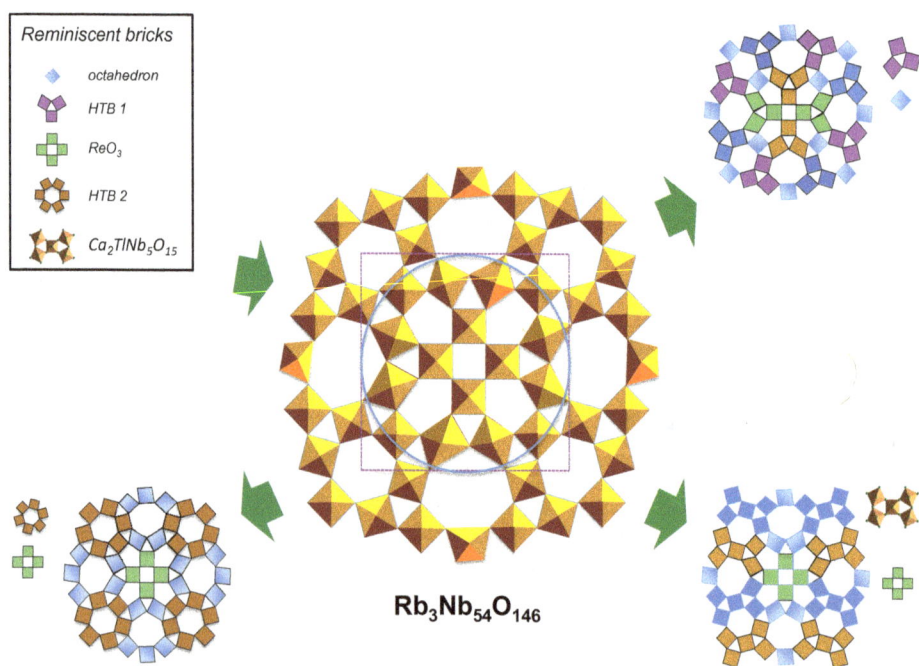

Fig. 6.22.   The various manners to describe the Gatehouse bronze. The central figure represents only one-fourth (purple dotted lines) of the pseudo-tetragonal cell. The TTB topology immediately appears inside the blue circle. The satellite figures illustrate three different ways for the description of this complex structure. The upper left part of the figure enumerates the different building blocks involved in this structure. Nevertheless, these building blocks are not sufficient for a complete reconstruction. The addition of supplementary single octahedra is always needed for this purpose.

a structural index sufficiently strong for observing heavy constraints in this structure, constraints which are insufficient for rendering this structure unstable.

6.5 $A_n[A'_{m-1}M_mX_{3m+1}]$ Structural Recurrences in Oxides (Raveau *et al.*, 1991)

Nothing better than a sandwich for introducing this paragraph!... Depending on his appetite, the reader will have many choices, from the simple one (thin slices of ham between two pieces of bread) to that with multiple layers alternating bread and several ingredients! However, Nature was the first producer of sandwiches, but this time, at the molecular scale... Humans came later!

The concept remains the same: stacking of different layers, with variable thicknesses (recurrence at the atomic scale) for generating a coherent solid (Fig. 6.23). The series presented below (mainly cuprates) correspond more to an increase of the thickness of the slices of bread of the sandwich than a variety of ingredients between them.

In the series in Fig. 6.23, each orange rectangle represents one oxide monolayer extracted from ReO_3 or perovskite structures. The stacking of m layers of orange blocks determines the skeleton of the final structure. Species are inserted between these multiple blocks. In the three families having the same general formula, it is both the chemical nature of A species and their positions with respect to blocks which will differentiate the families called Aurivillius, Dion–Jacobson and Ruddlesden–Popper phases.

6.5.1 *The Aurivillius phases*

In them, the signature characteristic of A is always a complex Bi_2O_2 monolayer. The bismuth titano-niobate Bi_3NbTiO_9, for example (Fig. 6.24) is built from the alternation of double perovskite layers (with Nb and Ti disordered on the octahedral sites and one Bi in dodecahedral ones), separated by Bi_2O_2 monolayers, described from squares of oxygen atoms capped alternatively above and below by Bi^{3+} ions.

This unusual pyramidal coordination is reserved for some cations (Bi^{3+}, Sb^{3+}, Tl^+...) which have in common the existence of active lone pairs on their external electronic layers; the single Bi_2O_2 layer is the signature of the Aurivillius phases. Instead of Bi_3NbTiO_9, it is therefore recommended to write the developed formula $[Bi_2O_2][BiNbTiO_7]$, which both distinguishes the two blocks and differentiates the two roles of bismuth. In the Aurivillius phases, note that the recurrence only concerns the number of perovskite layers (Fig. 6.25), the Bi_2O_2 monolayer being a constant within the series.

Intergrowth mechanisms of two structural units :
translation of perovskite blocks, m octahedra thick separated by different layers

(a)

Intergrowth mechanisms of cuprates : double recurrence

(b)

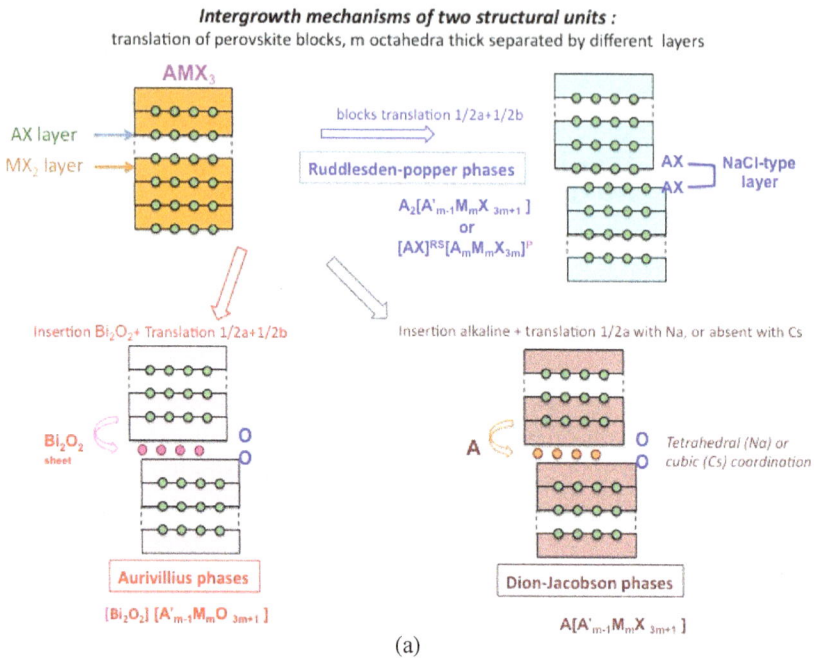

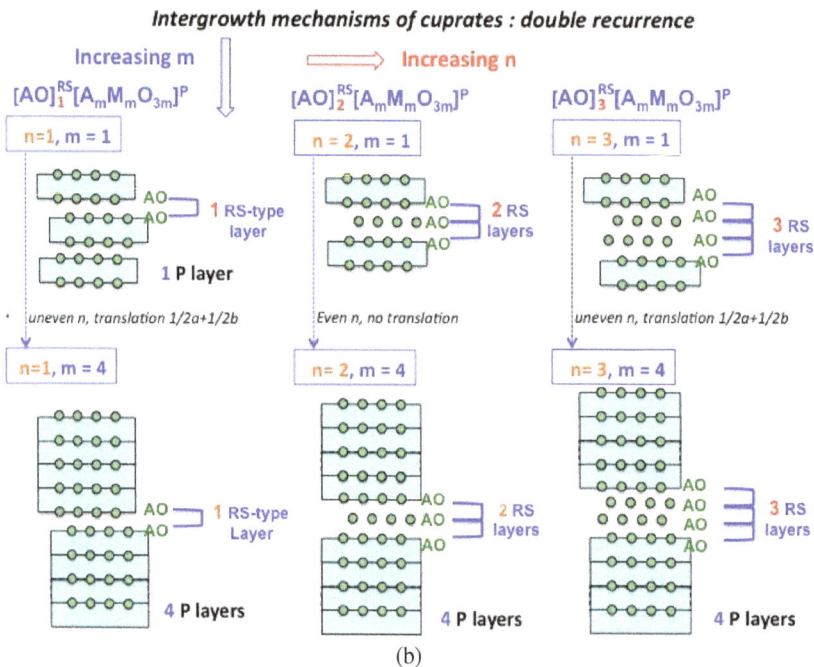

Fig. 6.23. Scheme of the principle of the formation of recurrent series of oxides from the general formula on the top of the figure. The nature and the position of layer A determines the three series whose names celebrate their inventors.

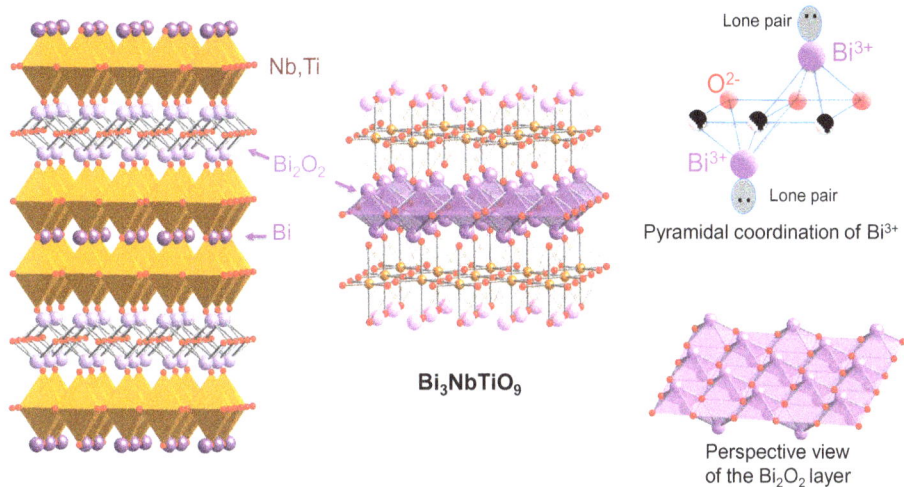

Fig. 6.24. Perspective view of Bi_3NbTiO_9 with its Bi_2O_2 monolayers (represented in several manners) inserted between the double layers $BiNbTiO_7$ and their Nb,Ti disorder. Lone pairs of bismuth appear as grey ellipses.

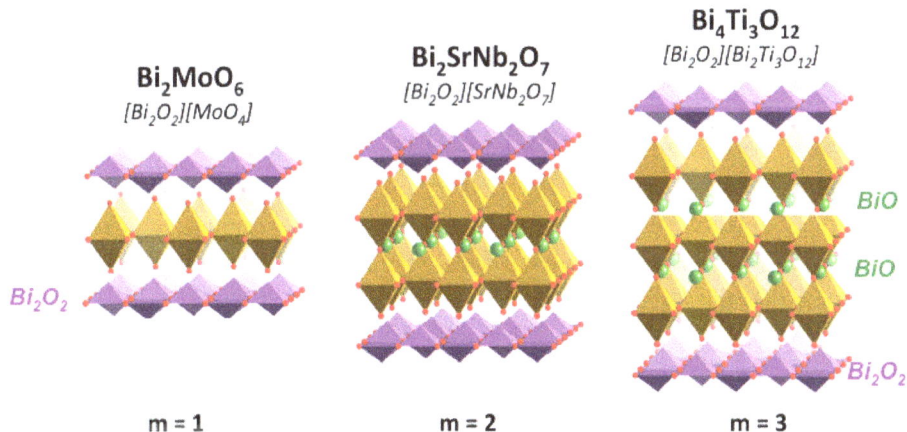

Fig. 6.25. The structural recurrence in the Aurivillius phases, due to the regular increase of the number of perovskite layers (in orange), here in the range $1 \leq m \leq 3$.

6.5.2 *The Dion–Jacobson phases*

They are deduced from the Aurivillius phases simply by replacing the Bi_2O_2 layers by monolayers of alkaline ions (Fig. 6.26), between the perovskite layers. The dodeca-hedral sites within the multilayers are often occupied by rare-earth trivalent ions.

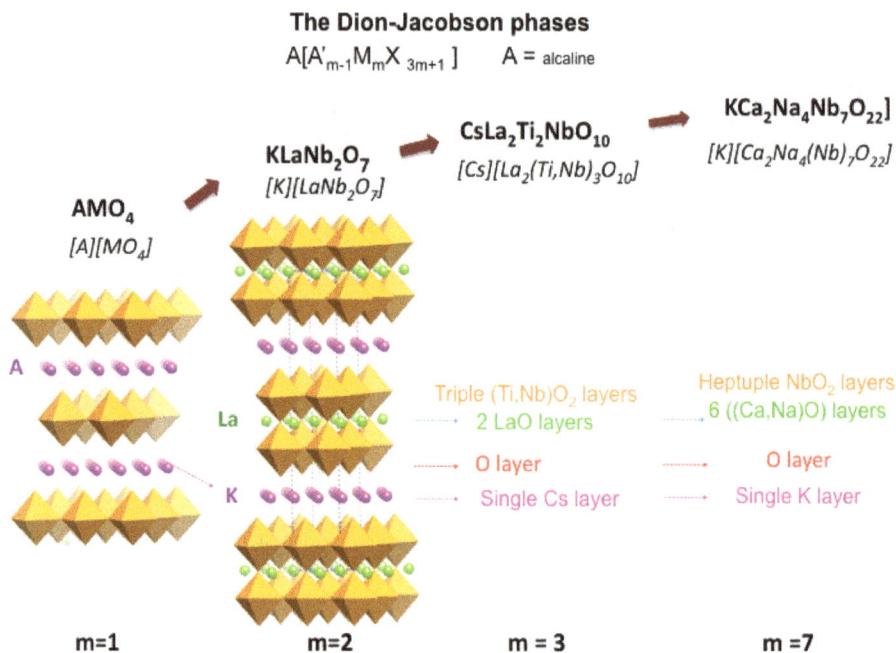

The Dion-Jacobson phases

$A[A'_{m-1}M_mX_{3m+1}]$ A = alcaline

KLaNb$_2$O$_7$
$[K][LaNb_2O_7]$

CsLa$_2$Ti$_2$NbO$_{10}$
$[Cs][La_2(Ti,Nb)_3O_{10}]$

KCa$_2$Na$_4$Nb$_7$O$_{22}$]
$[K][Ca_2Na_4(Nb)_7O_{22}]$

AMO$_4$
$[A][MO_4]$

A

La

K

Triple (Ti,Nb)O$_2$ layers
2 LaO layers

O layer

Single Cs layer

Heptuple NbO$_2$ layers
6 ((Ca,Na)O) layers

O layer

Single K layer

m=1   m=2   m = 3   m =7

Fig. 6.26.   Illustration of the structural recurrence in the Dion–Jacobson phases.

6.5.3 *The Ruddlesden–Popper phases*

Their originality comes from the particular localization of A cations. Instead of just insertions between the multilayers of the perovskite type, they are integrated at the surface of the latter, between four oxygen atoms of four different octahedra. Such an arrangement creates rocksalt AO planes at the surface of the multilayers. This situation means that, instead of having only one A cation in the general formula, there are now two (Fig. 6.27). Despite that, the law of recurrence is the same as for the other phases.

Within the perovskite-type multilayers, all these phases host a lot of cations with different charges, ordered or disordered, and create extremely rich chemical families.

6.5.4 *The structural recurrence in cuprates*
     (Raveau et al., 1991)

The structural adaptability of this new family is much larger than that for the phases described above. The latter had almost in common the invariance of the monolayers inserted between the blocks. One of the reasons of the adaptability of cuprates is the variety of coordinations that copper ions adopt, from the two-fold

The Ruddlesden-Popper phases

$[AO]^{RS}[A'_m M_m X_{3m}]^P$,   A = alcaline earth

Sr_2RuO_4
$[SrO]^{RS}[SrRuO_3]^P$

$Sr_3Ru_2O_7$
$[SrO]^{RS}[Sr_2Ru_2O_6]^P$

$Sr_4Ru_3O_{10}$
$[SrO]^{RS}[Sr_3Ru_3O_9]^P$

A' = Sr

A' = Sr

Sr

Sr

Sr

Sr

Sr

Double perovskite block

triple perovskite block

Two shifted Sr-O planes form 1 NaCl-type layer

Two shifted Sr-O planes form 1 NaCl-type layer

m = 1   m=2   m =3

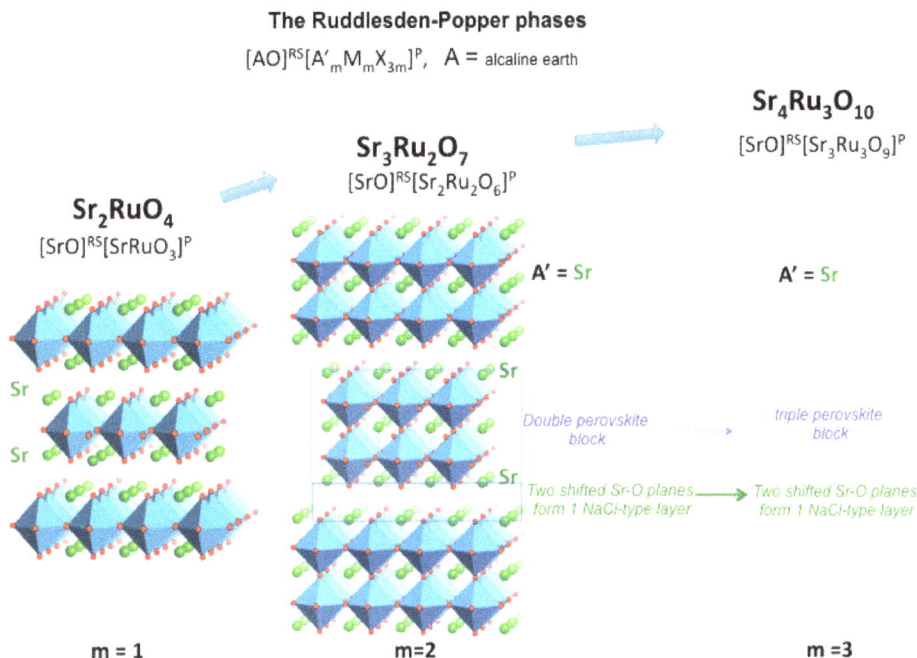

Fig. 6.27.   Illustration of the structural recurrence in the Ruddlesden–Popper phases.

for Cu^+ to the three for Cu^{2+}: square planar (coordination number C.N.: IV), square-based pyramidal (C.N. V) and, of course, octahedral. Another reason is their unprecedented faculty to accomodate ordered anionic vacancies in the structures.

This aspect was initiated in Chapter 3 on rather complex structures, but it exists even for simpler ones. Return to La_2CuO_4, isotypic with the Sr_2RuO_4 structure of Fig. 6.27 above. Replacing La^{3+} in La_2CuO_4 by Sr^{2+} leads, for reasons of electro-neutrality, to the elimination of one-quarter of the initial oxygen atoms and a correlative creation of vacancies. In the resulting solid Sr_2CuO_3, Cu^{2+} has lost its initial octahedral coordination to the benefit of a square planar one, despite the fact that the global topology is retained when vacancies are taken into account (Fig. 6.28).

In terms of structural recurrences, the La_2CuO_4 structure is a good starting case. Note first that La_2CuO_4 is a Ruddlesden–Popper phase because Sr^{2+} ions are at the surface of the perovskite-like monolayer. The space formed by the two opposite interfacial SrO layers, is reminiscent of a NaCl layer. Therefore, La_2CuO_4 represents an **intergrowth** between perovskite and NaCl-type layers. An intelligent chemist can think about the increase of the thicknesses of the two blocks: n NaCl layers and m perovskite ones. It is the case of $TlBa_2CuO_5$ and $Bi_2Sr_2CuO_6$ which can be written as $[TlBaO_2][BaCuO_3]$ ($n = 2$; $m = 1$) and $[Bi_2SrO_3][SrCuO_3]$ ($n = 3$; $m = 1$) for being more explicit in terms of

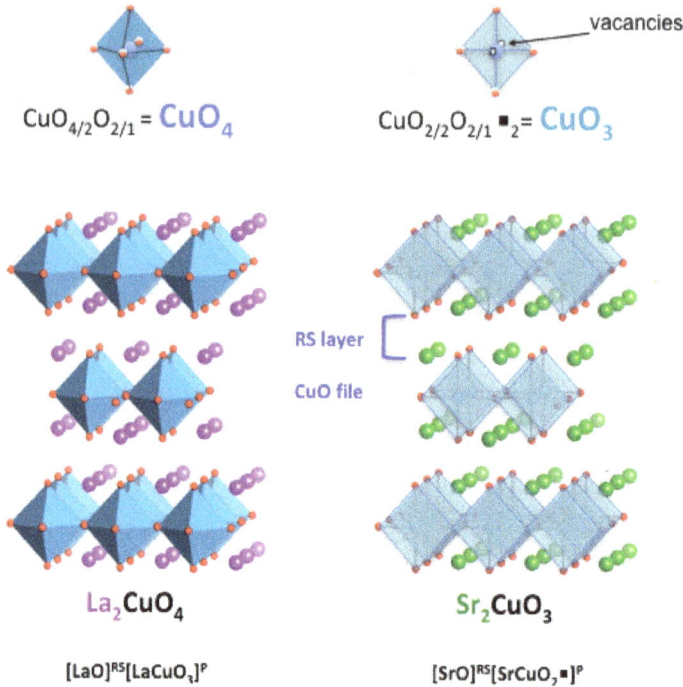

Fig. 6.28.   Creation of oxygen vacancies in the oxygen subnetwork by replacing trivalent La^{3+} by Cu^{2+} in La_2CuO_4. Even if the global topology remains the same, the coordination of copper (always 2+) has decreased from octahedral to square planar.

differentiated blocks (Fig. 6.29). This type of formula implies first that layers of oxide ions, associated with cations, will be present in the inter-blocks layers. This is the first step in the structural recurrence in cuprates: the progressive increase of NaCl blocks. They are highlighted in yellow boxes in Figs. 6.29–6.32. It is worthy to note that the Bi_2O_2 layers are of the NaCl-type in the following family, which implies a local difference for these layers toward those existing in the Aurivillius phases.

The interlayer space is now occupied in diverse manners. The second structural recurrence occurs this time by playing on the thickness of the perovskite layers, as it was done for Aurivillius phases (see Fig. 6.25). For reasons of stability, calcium ions occupy the cuboctahedra within the perovskite blocks, strontium being on the surface of these blocks. The resulting formula of recurrence are $[AA'O_2]^{RS}[A''_mM_mO_{3m}]^P$ if A is thallium with its Tl-O monolayer to form 2 rocksalt-type layers and $[A_2A'O_3]^{RS}[A'A''_{m-1}M_mO_{3m}]^P$ if A is Bi with its double layers to form 3 rocksalt-type layers. The general formula outlines the thickness of both the rocksalt and perovskite blocks $[AO]_n^{RS}[$ A'MO3$]_m^P$ (with A= Tl, Bi, Pb, Hg, RE, Sr.. and A'= Ba, Sr, Ca...); each of the numerous members of families are noted by four numbers *n, 2, m-1*, and *m* (Figs. 6.30–6.32).

$$[AO]^{RS}_n[A' MX_3]^P_1$$

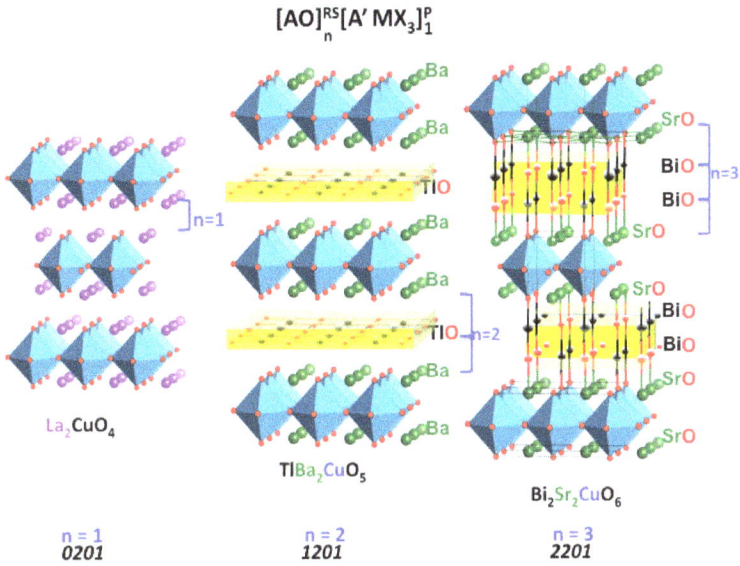

n = 1
0201

n = 2
1201

n = 3
2201

Fig. 6.29. Progressive filling of the space between the monolayers A'_2CuO_4 corresponding to the sequence $[A'_2CuO_4] \rightarrow [TlO][A'_2CuO_4] \rightarrow [Bi_2O_2][A'_2CuO_4]$ (A' = Sr, Ba, La). The [Tl–O] and $[Bi_2O_2]$ type NaCl blocks are highlighted as yellow boxes.

$$[AO]^{RS}_n[A' MX_3]^P_m$$

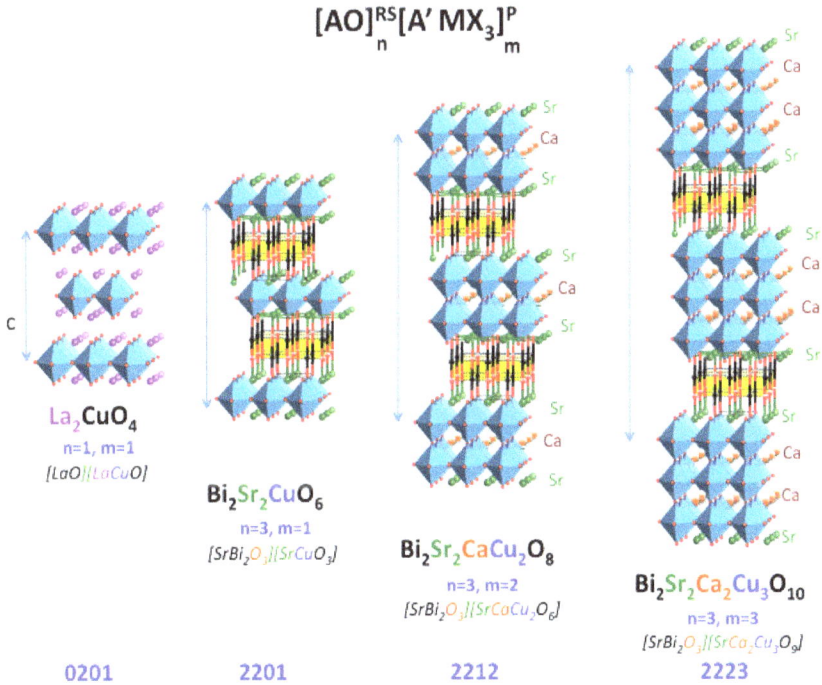

La_2CuO_4
n=1, m=1
[LaO]/[LaCuO]

$Bi_2Sr_2CuO_6$
n=3, m=1
[SrBi₂O₃]/[SrCuO₃]

$Bi_2Sr_2CaCu_2O_8$
n=3, m=2
[SrBi₂O₃]/[SrCaCu₂O₆]

$Bi_2Sr_2Ca_2Cu_3O_{10}$
n=3, m=3
[SrBi₂O₃]/[SrCa₂Cu₃O₉]

0201   2201   2212   2223

Fig. 6.30. Recurrence affecting the thickness of perovskite blocks, maintaining the same species (here (Bi_2O_2) layers in the inter-blocks space. For each solid, the developed structural formula is noted below the chemical formula. The representation of the $[Bi_2O_2]$ type NaCl blocks are highlighted as yellow boxes, as in Fig. 6.29.

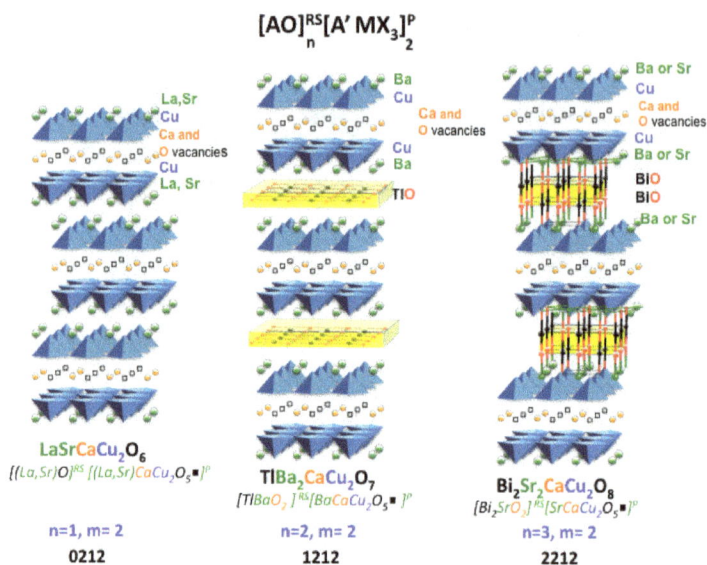

Fig. 6.31. Recurrence affecting the double perovskite blocks with vacancies, by playing on the chemical nature of the species in the inter-blocks. The representation of the [Tl–O] and [Bi_2O_2] type NaCl blocks is highlighted by yellow boxes, as in Fig. 6.29.

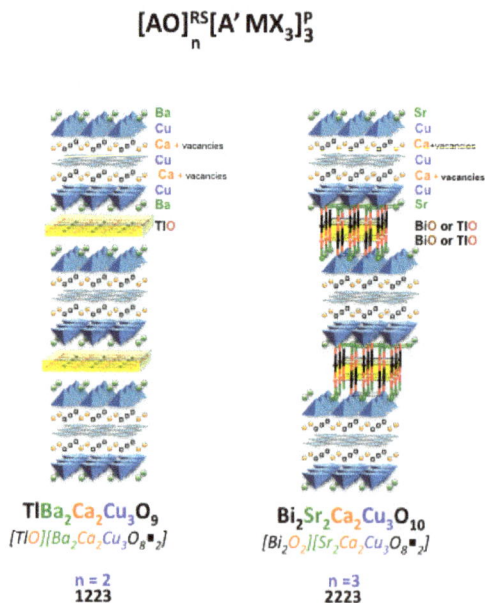

Fig. 6.32. Recurrence affecting the triple perovskite blocks with vacancies, by playing on the chemical nature of the species in the inter-blocks. The representation of the [Tl–O] and [Bi_2O_2] type NaCl blocks is highlighted by yellow boxes, as in Fig. 6.29.

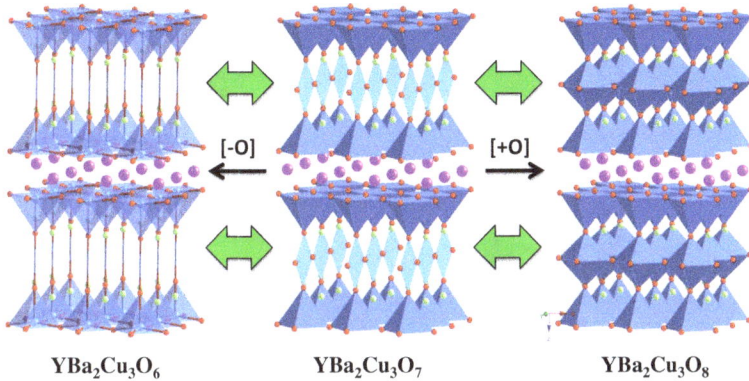

Fig. 6.33.   Evolution as a function of oxidation in the YBaCuO structure.

All these series correspond to the octahedral coordination for copper. With La_2CuO_4 and Sr_2CuO_3, it appears that, playing rationally on the chemical nature, the charge and the ratios between the cations occupying either the cuboctahedra of the perovskite blocks or the interlayer space, it was possible to create vacancies. The same strategy is always valid in the present series, whatever the thickness (Figs. 6.31 and 6.32). Only the axial oxygen atoms in the central blocks disappear and then create vacancies in their place.

The last example, with its triple perovskite layers, is particularly interesting. It is reminiscent of what happened for the famous YBaCuO (the supraconductor $YBa_2Cu_3O_7$) and its variations of configuration during its oxidation (Fig. 6.33). It was not unique and the above recurrence law shows that, despite its celebrity, this solid is just one (privileged) member of a very large structural family.

Even when solids present triple perovskite layers with always the same 1/2/3 ratio for Y/Ba/Cu (which explains the alternative 123 notation sometimes used for $YBa_2Cu_3O_x$), it is now clear that, in different chemical conditions, with judicious choices of nature and the size of counter-cations, it is possible to reach different local topological situations (Fig. 6.34).

These choices are strategic. Indeed, coming back to the 123 blocks of YBaCuO, another mixture of counter-cations leads to a different situation. The latter can serve as chemical scissors modifying the arrangements. It is the case of $Sr_6Nd_3Cu_6O_{17}$ (Fig. 6.35). The couple Sr/Nd cuts the triple blocks which were primitively infinite in two dimensions, in such a way that the infinite development occurs now in only one dimension, the other being restricted to two octahedral rows.

Finally, a supplementary recurrence can exist within the 123 blocks. It concerns the files of square planes. They are not always single files (Fig. 6.36). Look

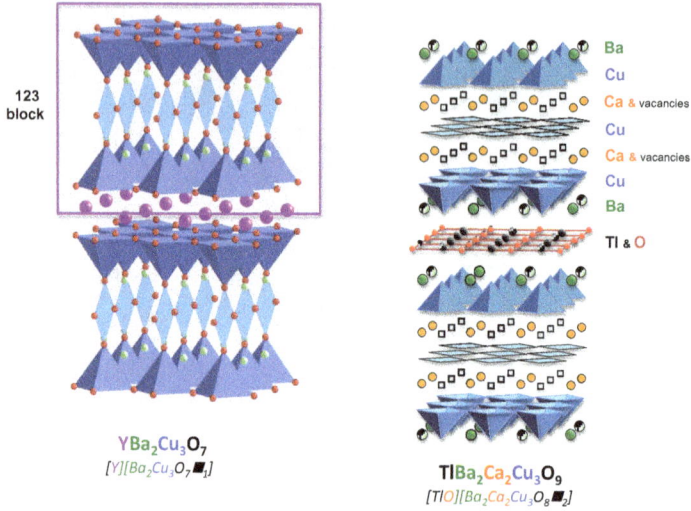

Fig. 6.34.   Different orders of the anionic vacancies in the triple perovskite layers as a function of the choice of the counter-cations.

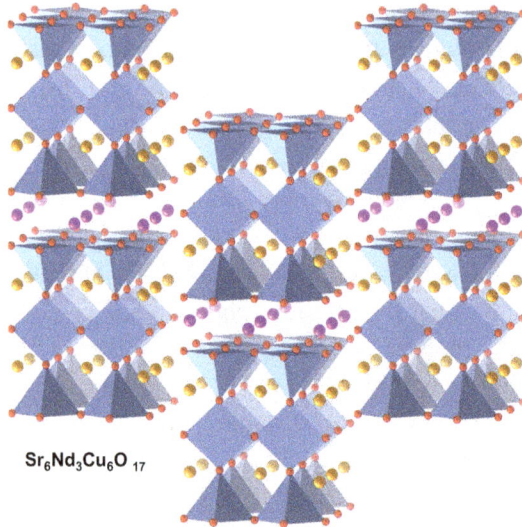

Fig. 6.35.   Cut induced by Sr/Nd couples of the blocks ($3 \times \infty \times \infty$) into ($3 \times 2 \times \infty$) blocks in $Sr_6Nd_3Cu_6O_{17}$.

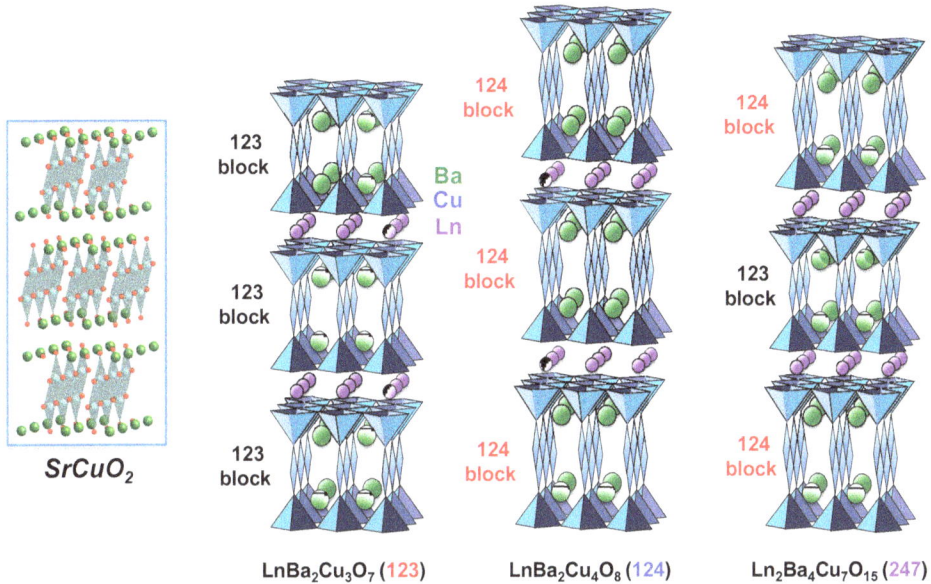

Fig. 6.36.   Diverses insertions of doubles chains of square planes in the matrix 123.

indeed at $SrCuO_2$. This structure exhibits double chains of edge-sharing square planes (insert of Fig. 6.36). The (partial or complete) substitution of single files by double files can be obtained just by a slight change, at the beginning of the synthesis, on the composition of the starting mixture. A complete substitution corresponds to $LnBa_2Cu_4O_8$; a partial one to $Ln_2Ba_4Cu_7O_{15}$ [Fig. 6.36 (right)].

It must be noted that $LnBa_2Cu_4O_8$ before being isolated as a pure phase, was observed six months before by high resolution electron microscopy as an impurity in a bad synthesis of 123.

All the numerous examples of cuprate structures detailed above started from cubic perovskite topology. But this topology is not the only one which gives rise to recurrent structure types. In the following, another family named **hexagonal perovskites** will be considered. Even though commonly used by crystal chemists, this term is ambiguous. These solids are not derived from the cubic perovskite. Topologically they are rather different. The main difference concerns the stacking of anionic layers. In cubic perovskites, it was a ABC stacking; in this new family, it is a AB one. This change in the stacking of similar layers probably explains why many crystal chemists created the term **hexagonal perovskites** to qualify a pseudo-relation which is considered as artificial by the author, despite keeping this term for the sake of simplicity.

6.6 Structural Recurrence in Hexagonal Perovskites

The left part of Fig. 6.37 recalls the different views of the cubic perovskite structure described in terms of a ABC packing of dense layers (see legend of Fig. 6.37). If only one of these triangular layers is considered and if one stacks two of them in such a way that the octahedra share faces one layer to the other, [Figs. 6.37(f)–6.37(g)], the structure of $CsNiF_3$ hexagonal is generated [Fig. 6.37(h)]. Its periodicity corresponds to two layers and is noted **2H** symbolizing both the periodicity (2) and the symmetry (H) of the corresponding cell. This represents the starting model for all the hexagonal perovskites, on which several forms of recurrence will be applied.

The building block for this first type of recurrence is the bioctahedron sharing a face (Fig. 6.38). After a translation of $(\frac{1}{3}a + \frac{2}{3}b + c)$ such bioctahedra can share vertices. The corresponding structure, existing for one of the polytypes $BaRuO_3$, has a periodicity equivalent to four layers and is noted 4H. One can also intercalate

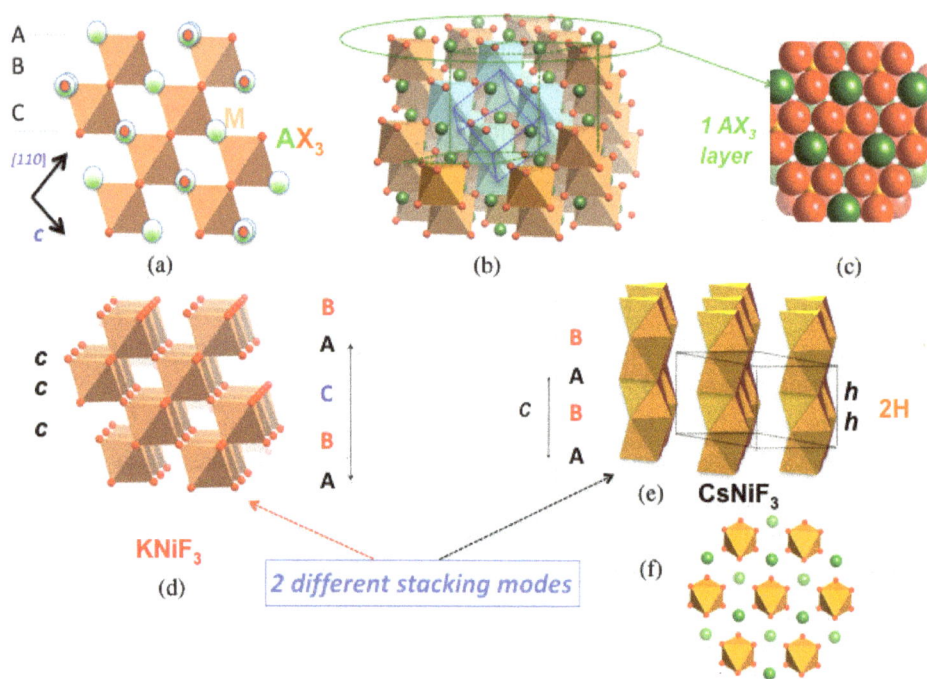

Fig. 6.37. Various manners for looking at the cubic perovskite; (a) classical view; (b) its perspective view along a direction roughly perpendicular to the [111] direction of the cubic cell; in it, appear the cubic (blue) and hexagonal (green) with their relative orientations. The AX_3 layer (A: green; X: red), highlighted by a green ellipse, is represented in projection on (c); Perspective view in terms of the stacking of dense layers in $KNiF_3$ (d) and in $CsNiF_3$ noted **2H** (e) with their two usual notations; (f) (001) projection of $CsNiF_3$.

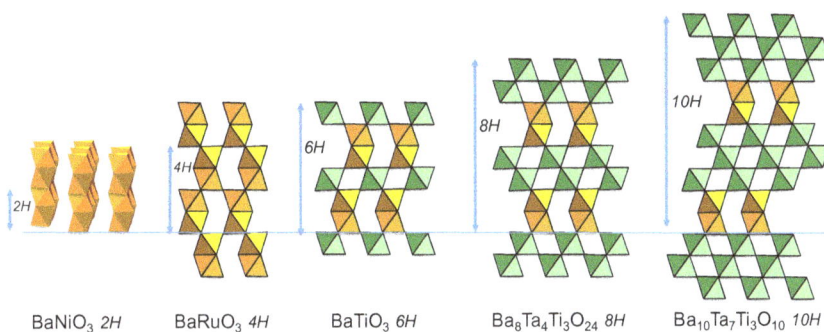

Fig. 6.38.   First type of recurrence by insertion of c_p layers between the bioctahedral ones.

single or multiple layers c_p of the perovskite type between each layer of bioctahedra. This leads to structures presenting 1 to n $(n \leq 3)c_p$ layers for separating the
layers of bioctahedra (Fig. 6.38) which increase the periodicity along the c axis of
the hexagonal cells.

However, in this large family, the bioctahedron is not privileged (Fig. 6.39).
Numerous structures are indeed based on trioctahedra, alone ($BaRuO_3$ 4H) or
intercalated by c_p monolayers ($Ba_5Nb_4O_{15}$ 12R). In the last case, the symmetry
becomes rhombohedral. These trioctahedra can even accept vacancies (see
$Ba_5Nb_4O_{15}$ 12R). A few cases exist with face-shared tetraoctahedra and even
octaoctahedra as building blocks. This leads to very large values of the c parameter
of the cell, for instance with one of the forms of $BaCrO_3$ (27R). This new family
can therefore be considered as a second type of recurrence: insertion of layers of
face-shared multioctahedra between c_p monolayers.

For a given chemical formula, these hexagonal perovskites present a strong
polytypism. It was already seen with some of the forms of barium ruthenate
$BaRuO_3$. This structural variety is increased by a size effect when one plays on the
replacement of Ba by the smaller Sr. In the phase diagram versus pressure in the
$Ba_{1-x}Sr_xRuO_3$ family, represented in Fig. 6.39, there successively appear, at
increasing pressures, the forms 9R, 4H and 6H. In this diagram, the white zones
correspond to a mixture of phases. Note finally that a large proportion of strontium
is needed for obtaining the cubic perovskite.

Beyond the perovskite topologies (hexagonal and cubic), recurrences on other
structure types also exist when the starting structures share edges. The $CdCl_2$ type,
studied in Chapter 3, is pertinent on this point (Fig. 6.40). The illustration is provided by silver nickelates. Starting from the NiO_2 layers, similar to those existing
in $CdCl_2$, the inserted Ag^+ form hexagonal nets between the NiO_2 blocks. A supplementary insertion of these ions gives the solid Ag_2NiO_2, with the formation of
Ag^+ double layers $[(Ag^+)_2]_\infty$, shifted in such a way that they form octahedral

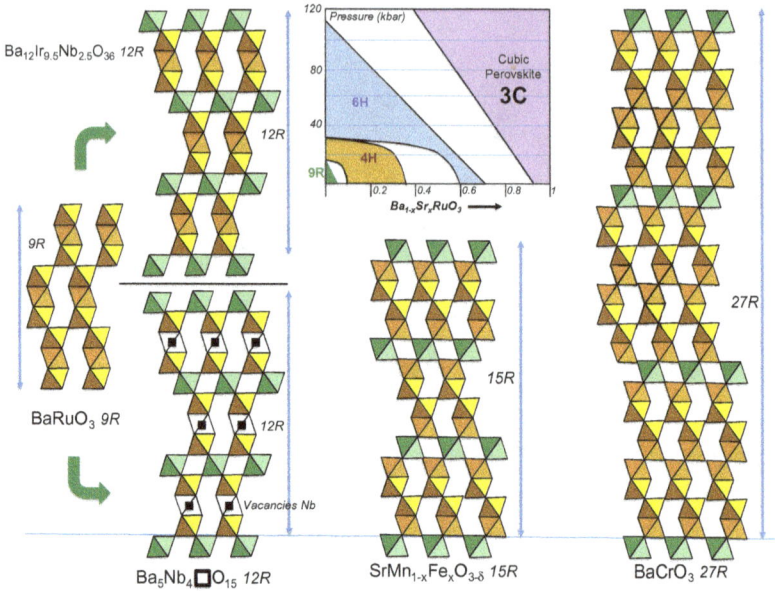

Fig. 6.39.   Second type of recurrence by insertion of multi-octahedral layers between c_p monolayers. The central diagram represents, as a function of the x proportion of strontium in the solid solution $Ba_{1-x}Sr_xRuO_3$, the structural evolutions induced by the application of pressure.

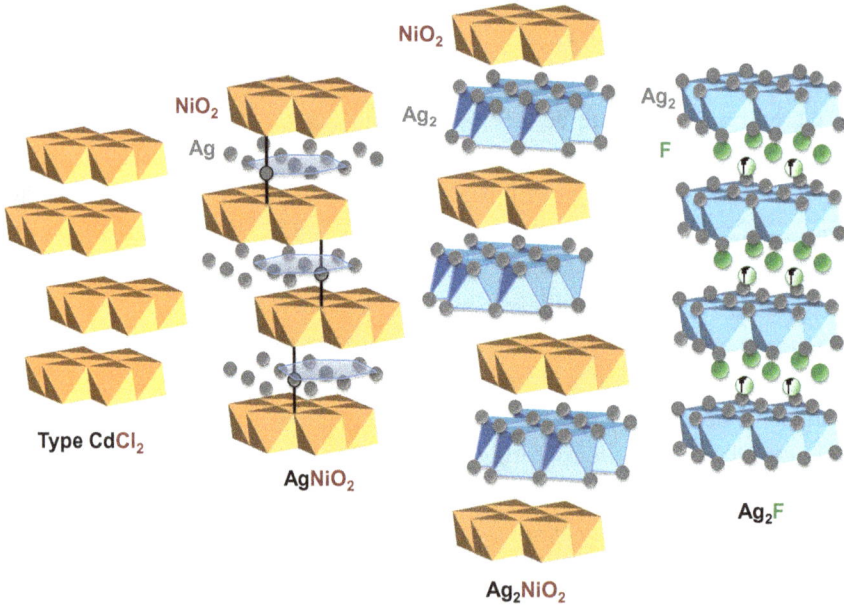

Fig. 6.40.   Recurrence by a progressive insertion of cationic layers within those of $CdCl_2$.

sheets; the vertices of these octahedra are occupied by silver ions; their center is empty. These $[(Ag^+)_2]_\infty$ layers are also found in the inverse structure of AgNiO$_2$: Ag$_2$F. More generally, this type of stacking is also found in series of manganites and cobaltites.

The discovery of structural recurrences is always satisfying for at least three reasons: (i) the elaboration of the *genealogy* for a structure recently discovered. It is important to consider that this new structure is not an isolated case, but belongs to a large family with its differences, its specificities; beyond this *genealogist* work, they enhance the admiration that the crystal chemist feels in front of the extraordinary faculties of construction of new solids from the fundamental molecular bricks which, mysteriously, are often derived, despite the scale differences, from the Platonic and Archimedian polyhedra; (iii) by making an intelligent use of the building blocks rules, overcome the step devoted to the careful analysis of structures by the crystal chemists for reaching another essential facet of the job of a chemist: the creation of new architectures on a rational basis, instead of using the traditional trials and errors method. The chemist is also an innovative creator.

From these recent examples, Chapter 7 will show that this creative approach is possible but, before that, there is one point that must be detailed: the intergrowth phenomena. They could have been evoked in another paragraph on structural recurrences, but the latter involves the association of finite objects with the same dimension with recurrences in simples ratios. The concept of intergrowth, explained below from two examples, is much more general because it does not imply a strict periodicity between two topologies.

6.7 Other Mechanisms of Intergrowth

For understanding them, it is necessary to take into account the *magic* operator: the 60° angle present in different bricks. It obviously exists as perfect in the HTB hexagonal tungsten bronze, (Fig. 6.40), but it also exists in tilted perovskites. Indeed, take a 2×2 perovskite block. In one plane, the tilting is maximum when, in a plane, the initial square is transformed into a regular lozenge. This coincidence immediately allows to conclude that HTB and a tilted perovskite (or ReO$_3$) structure are compatible for creating an interface associating the two topologies. For being clearer, one must come back to these well known topologies, particularly the HTB structure which was previously described in terms of trimeric blocks. However, intergrowth can be considered also as the association of linear chains of the tilted ReO$_3$ type (in orange) with branched chains with satellite octahedra (in blue) (Fig. 6.41). The latter, in this description, will represent another kind of HTB building block. The connection between the two

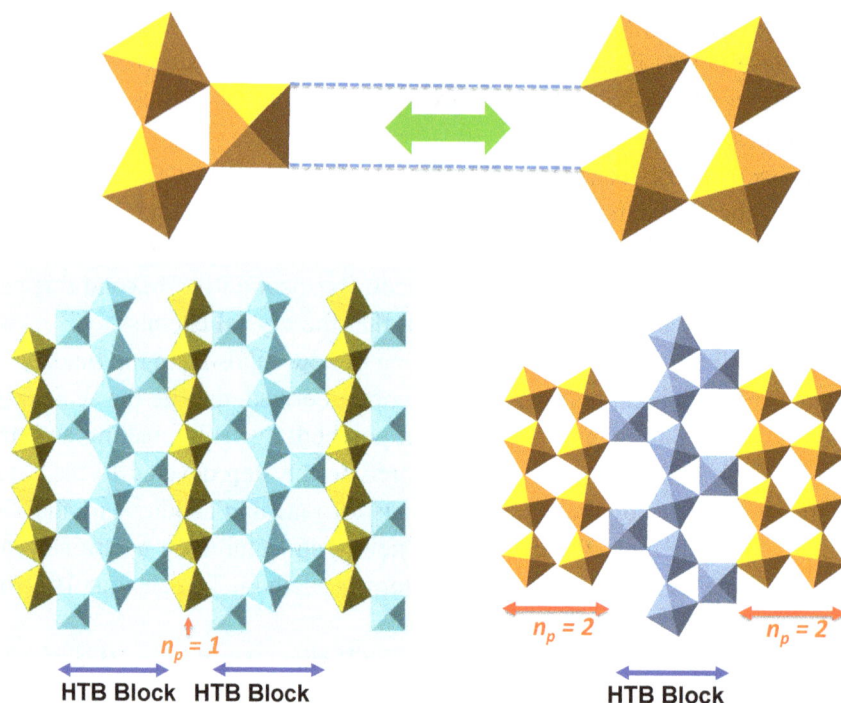

Fig. 6.41. (Top) Compatibility between a triangular trimer of octahedra and a tilted square of octahedra; (bottom left) a new description of HTB corresponding to the association of the new HTB blocks (in blue) with single chains of tilted ReO_3 noted n_p (in orange); (bottom right) a first term of recurrence playing on the same HTB block with two corner-shared chains n_p of the ReO_3 type with $n_p = 2$.

types of chains is possible because the distance between the two oxygen atoms of the ReO_3 tilted chain is exactly the same as their distance within the edge of an octahedron of the HTB block [Fig. 6.41 (top)]. These oxygen atoms of the two species can be fused and then the connection is ensured through this inter-growth plane in order to regenerate the HTB structure when a ratio 1:1 exists between the two types of chains. This opens the way for the creation of new family of solids by an intergrowth mechanism between the two structures ReO_3 and HTB.

The recurrence uses only one HTB block and the incrementation plays on the number of n_p connected chains between these blocks. Figures 6.41 and 6.42 illustrate the ordered cases corresponding to $n_p = 2$ and 6. In the following, each member of this new series will be noted by 1,n (see the insert of Fig. 6.42).

For the moment, as the different blocks remain identical to themselves, there is no difference between intergrowth and the above structural recurrences but it is rather rare. More often, looking at the structural organizations using electron

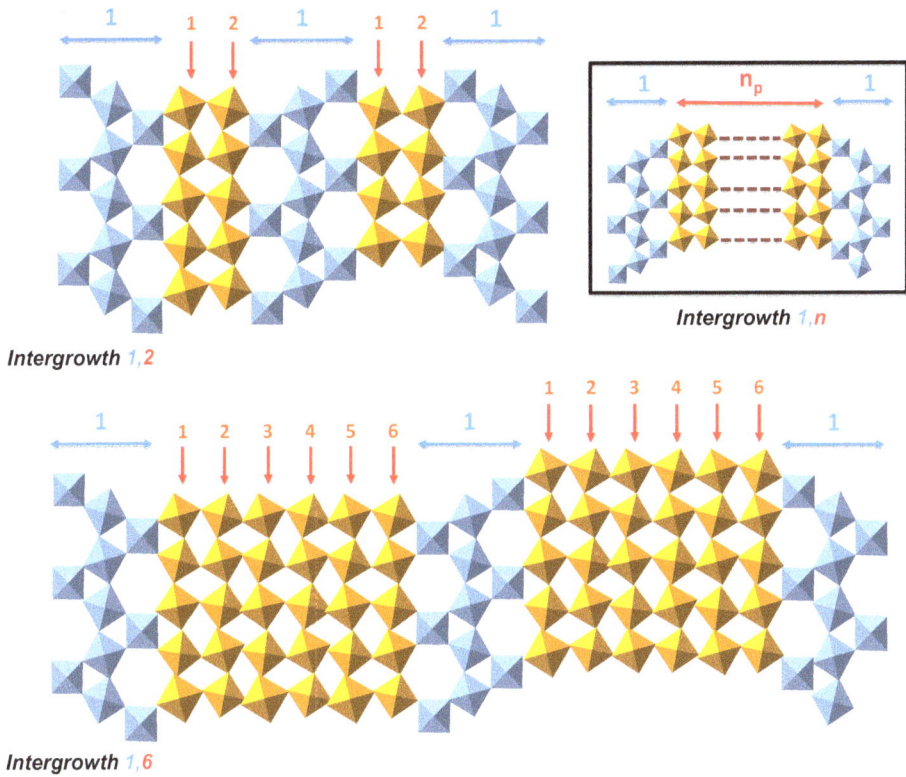

Intergrowth 1,2

Intergrowth 1,n

Intergrowth 1,6

Fig. 6.42. The examples of 1,2 and 1,6 periodic intergrowths between one HTB block and a variable number of tilted chains of the ReO_3 type.

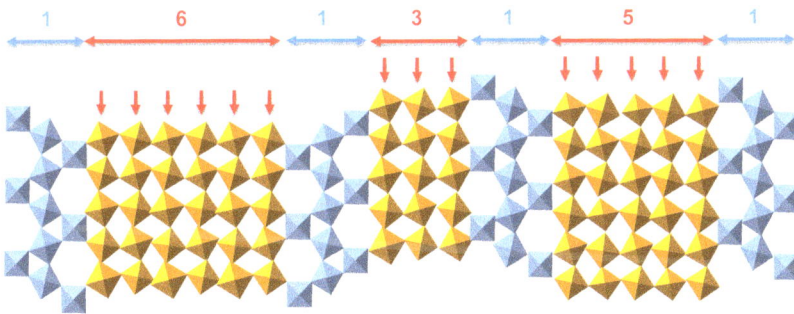

Fig. 6.43. Example of an inhomogeneous intergrowth with variable values of n_p in a sample.

microscopy, it is possible to locally distinguish within the same sample, a variable number of n_p values, whereas the global topology seems to remain unaffected. Recurrence disappears but intergrowth is always present (Fig. 6.43)... (Khilborg *et al.*, 1988).

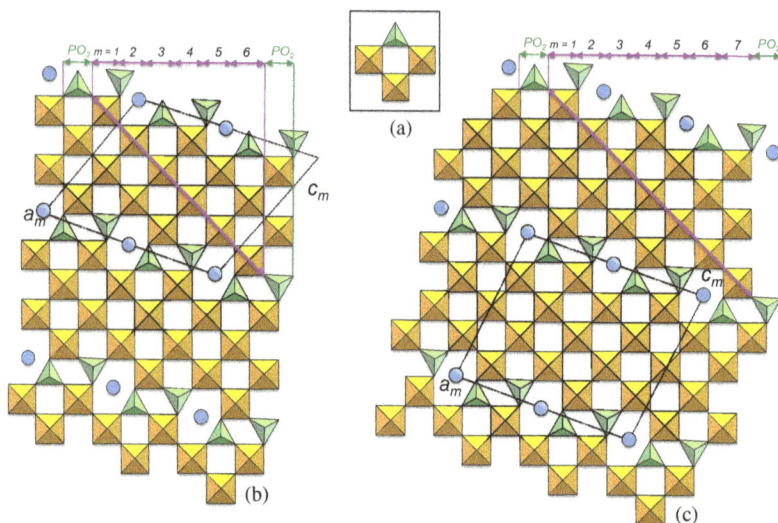

Fig. 6.44. (a) Scheme of compatibility for the connection by corners between the faces of tetrahedra and non-tilted ReO_3 blocks; (b–c) two examples of recurrence on the WO_3 blocks in tungsten bronze phosphates for two values of m ($m = 6$ in (b) and $m = 7$ in (c)). In each case, the blue circles represent the A ions, always located in hexagonal tunnels; the monoclinic cells are also represented.

The second example concerns monophosphate tungsten bronzes (MPTB) (Labbé, 1986) (Fig. 6.44). Shown above is the compatibility between triangular trimers and tilted squares for octahedra. Due to the 60° angle of the basal triangular faces of tetrahedra, the same situation also exists for connecting ReO_3 blocks (this time not tilted!) with tetrahedra for the generation of solids with a long distance ordering [Fig. 6.44(a)]. The textbook example for this connection is provided by the monophosphate tungsten bronzes family. Its general formula is $A_x(PO_2)_2(WO_3)_{2m}$, A being an alkaline ion (K, Rb, Tl) or baryum. x is always close or equal to 1 and m is an integer.

If m was constant in the whole sample, these solids would provide a new example of recurrence but, as it was the case for the above intergrowth bronzes, electron microscopy often shows inhomogenities in composition, with m randomly variable.

6.8 Structural Evolutions Under Pressure

This evolution was briefly recalled with the mechanism of transformation of the rutile form of TiO_2 into its high-pressure form TiO_2–II (or HP). However, it must be noted that the progressive increase of pressure can lead to many polymorphs and not only one, as it could be thought from the TiO_2 example. Our earth itself probably provides the best example of these successive transformations. Geologists have shown that the cubic mineral perovskite exists in very large amounts on earth, besides the prevailing silicates and alumino-silicates. A question however arises: how is it formed? Is it naturally and immediately formed in the conditions in which we live? Or is it the result of the transformation of other minerals existing under the mantle, and which are submitted to high temperatures and pressures? For getting an answer to this question, numerous geologists and mineralogists have performed studies since a long time, looking at the structural evolution under pressure of a lot of natural minerals in order to establish filiations between apparently different structures (Mitchell, 2002).

The studies on cadmium germanate $CdGeO_3$ have provided much information. Its phase diagram indeed reveals the complexity of its phase transformations (Fig. 6.45). Both cadmium and germanium ions have several interesting characteristics in terms of various coordinations and various ionic radii. They can exhibit several coordination numbers: from V to VI for germanium; generally six to eight, (even 12 for cadmium), and have very different ionic radii, function of their coordinations (from 0.92 Å to 1.45 Å for Cd when coordination passes from IV to

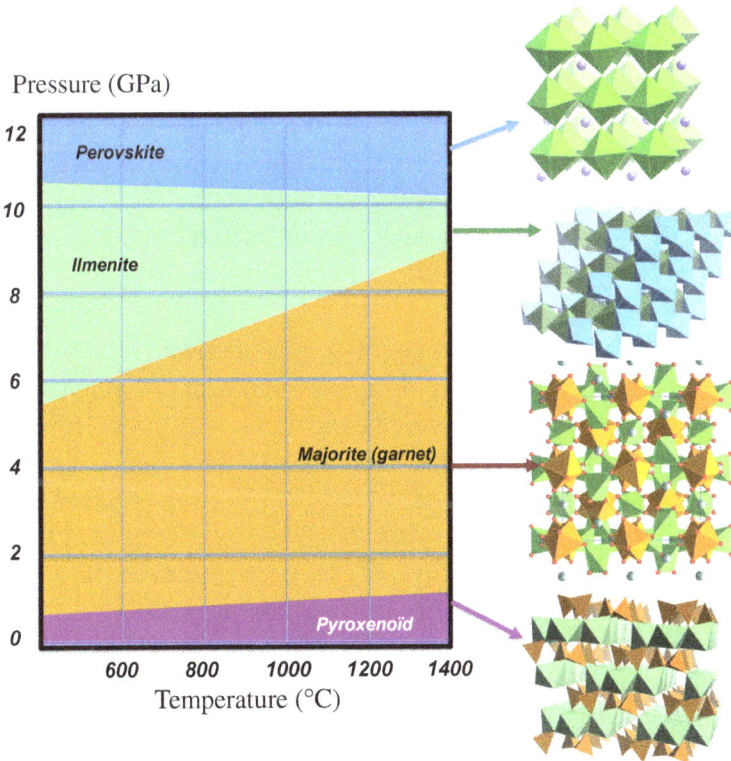

Fig. 6.45.   (P,T) diagram of $CdGeO_3$ with the corresponding structures.

XII; from 0.53 when it adopts a tetrahedral coordination to 0.67 Å when Ge is surrounded by an octahedron of six oxygen atoms.

The increase of pressure for a given temperature leads to the apparition of four distinct structures:

— The pyroxenoïd in which germanium forms corner-shared chains of tetrahedra connected to the other by layers of edge-shared cadmium octahedra;
— The majorite (with the garnet type). The general formula of this complex cubic structure is $^{VIII}A_2{}^{VI}B_2{}^{IV}C_3O_{12}$, in which the exponents indicate the coordination of each site. Cadmium is distributed on the three A, B and C sites; germanium resides only on the octahedral B sites and on the tetrahedral C sites;
— The ilmenite in which both Cd and Ge are in octahedra. This structure was already described in Chapter 4, Fig. 4.11. Deriving from the perovskite, the octahedra exhibit alternated 30° rotations around the [111] direction;
— A tilted orthorhombic perovskite.

Even if intuitive, it seems paradoxal, a first feature emerges: increasing pressures increase the coordination of cations. A second piece of information concerns the preponderant influence of the size of the divalent ion. Indeed, in a similar study related to $CaGeO_3$ (1.14 Å $\le r_{Ca^{2+}} \le 1.48$ Å depending on the coordination), the Ilmenite is no longer present, calcium being too large for adopting an octahedral coordination in this structure. It prefers a dodecahedral site in the perovskite.

Another effect of pressure concerns the distortion on a given structure, keeping the topology. It is proved on two well-known structures: ReO_3 and $BaTiO_3$ (Fig. 6.46).

As a first step, pressure favors a tilting of the octahedra in ReO_3. It is worthy to note here the antagonistic effect of pressure and temperature on the structural transitions: if a pressure increase results also in the magnitude increase of tilts, increasing temperatures decrease and even suppress them. It was already seen for FeF_3 and its transition rhombohedral-cubic at 395°C. Moreover, at high pressure, the tiltings in ReO_3 become so important that the symmetry changes from cubic to hexagonal while keeping the topology, but the density of the phase is increased.

In open structures, the pressures of transition are rather weak. With dense structures, higher pressures are necessary because the interstices of the structures are filled. The transitions are more difficult to perform. It is particularly clear when comparing ReO_3 and $BaTiO_3$. The transition of the latter needs the actions of both high pressure and temperature, and leads to a reorganization involving the

ReO₃

Cubic Pm3m   Cubic Im3   Hexagonal P622

1.5 GPa
300 K

9 GPa
300 C

BaTiO₃

12 GPa
600 C

Tetragonal P4mm   Hexagonal P6₃/mmc

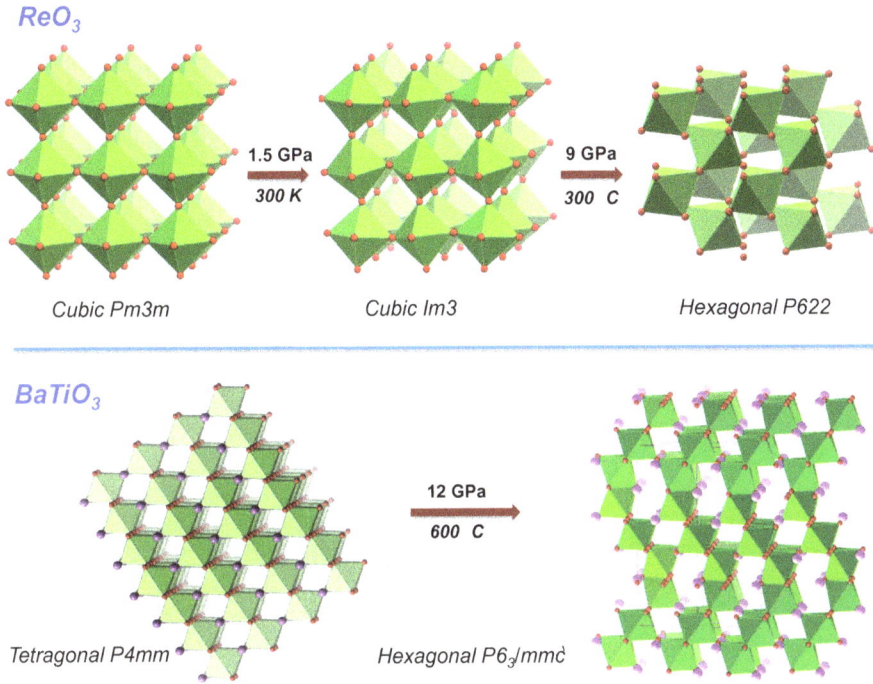

Fig. 6.46. Evolution of the ReO₃ and BaTiO₃ structures versus pressure.

glide of octahedral layers. A corundum form then appears, with the onset of face-shared octahedra.

At the end of this last paragraph, the reader has now all the trumps in hand for becoming a crystal chemist. All the analytic aspects are now at his disposal. It is time now to jump to another step and create new solids on a rational and often predictive basis. The last chapter opens some avenues on this specific point.

Chapter 7

Crystal Chemistry? A Tool for Creation... or... The New Architects of Matter

Tu m'as donné ta boue, et j'en ai fait de l'or.

Charles Baudelaire

The trial and error method has been, since a long time, the favorite tool of chemists for the creation of new solids. It proved its efficiency, remains valid in the immense majority of cases and deserves respect. However, from experience, the chemist only analyzes *a posteriori* the characteristics, including structural, of the products he synthesizes and transforms. His sight on the results allows him, after structural studies, to extract tendencies, evolutions often from geometric criteria for establishing classifications in relation with future properties. It is therefore a *deductive* approach. Must chemists stay there? Is it possible to find a new approach, more inductive, which starts *a priori* from a given topology, imagines and creates new matter whose structure presents analogies with the original topology? It was done in music. Starting from a simple melody, a composer can write variations which enhances (often with virtuosity) this initial melodic line. Mozart, Liszt have done it beautifully in the past. In the same way, can chemists realize variations from well known basic structures? It was a challenge, and only good crystal chemists could solve the problem by using a strategy inverse to that followed in Chapter 6. Indeed, structural recurrences come from a deductive approach.

Surprisingly, this challenge was taken up at the same time, in two different countries and published in the same journal (Journal of Solid State Chemistry), in the same issue (n° 1, volume 152).... In the history of Science such coincidences occasionally occur. The same idea, eventually expressed in different manners, nucleates simultaneously in different places. The most famous example in chemistry was the simultaneous discovery of fluorides of rare gases, both in Germany by Rudolf Hoppe and in USA by Neil Bartlett. These gases had before the reputation to be completely inert during chemical reactions.

This coincidence once more occurred in crystal chemistry, this time with the author of this book and Mike O'Keeffe, from Arizona State University. The first developed the *scale chemistry* concept; the second that of *augmented nets*. Even if the starting object seems different (bricks for the first, nets for the second), their concepts have in common the idea of structural homothety and enhancement. They often led to giant structures, *a priori* complex, but easily readable. These concepts will be first described before illustrating the creative power that they can both generate.

7.1 The Scale Up Concept: (Férey, 2000)

This concept offers another look at known structures, considered this time as the assembly of secondary building units (SBUs), whatever their size, their symmetries and formulas. It is a topological approach which ignores in a first step the chemical compositions and only focuses on the ways of connecting between these variable bricks.

Take for example a primitive cubic cell (P). It exists for polonium. Its building unit is the Po atom itself. ReO_3 has also a P lattice, but the building unit becomes the octahedron, while keeping the same topology. Going farther, in the Linde A zeolite (a complex sodium silico-aluminate), which also has a P lattice, the interesting building unit can be seen as an octahedron at the vertices of which are double-four rings D4R in which Al^{3+} and Si^{4+} alternate (Fig. 7.1).

This first relation seems to indicate — in the hypothesis of a real existence of the brick during the reaction — an invariance of the global topology for the different but isomorphic SBUs, despite their differences in size which depend on the complexity of each SBU. Only the shape and the size of the interstices are affected by the change of scale.

This approach has another advantage. It proves that, for the description of structures, the choice of the classical polyhedra as a tool of description is too restrictive. Beyond these polyhedra, well characterized atoms aggregate, polyatomic entities can be used for explaining a structure.

The concept remains valid for other types of topologies, proved for centered and face centered lattices (Figs. 7.2 and 7.3).

This means that it could be possible to describe simple structures from the structural topology of metals. But is it always applicable when several types of SBU are used? Clearly yes. Figure 7.4 is a proof with the assembly of cationic and anionic SBUs, from NaCl with its two types of octahedra $NaCl_6$ and $ClNa_6$ up to cobalt sulfide. In the latter, cobalt is involved in two different bricks: a classical octahedron and an octamer, formed by eight edge-shared tetrahedra already encountered during the study of face-centered cubic metals. Whatever the nature of the elements, their coordinations and the connection modes of the initial polyhedra forming the block, the topology (here *fcc*) is retained.

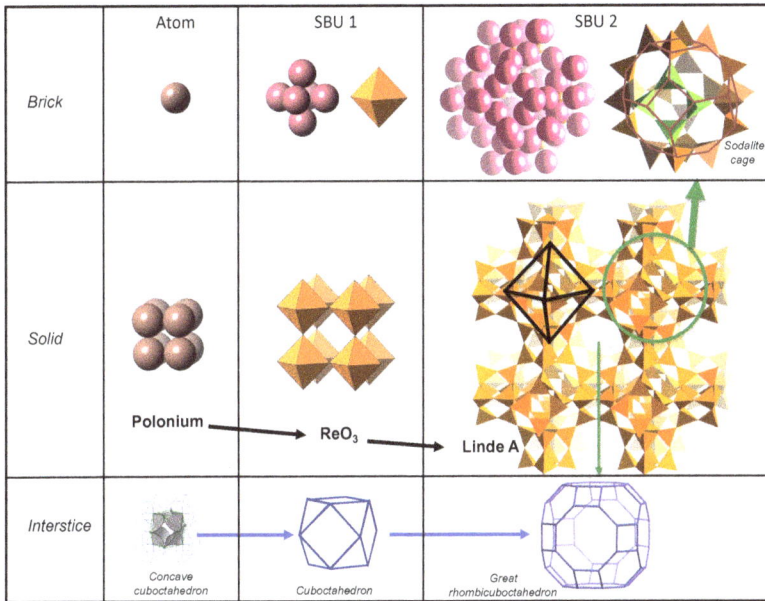

Fig. 7.1. Scale chemistry in cubic P lattices. Whatever the size of the building block (represented in the upper row in terms of polyhedra and of real size of atoms), the topology of all these solids remains invariant, type ReO_3 (middle row). The evolution of the shape of the interstices with the size of the SBU appears in the bottom row.

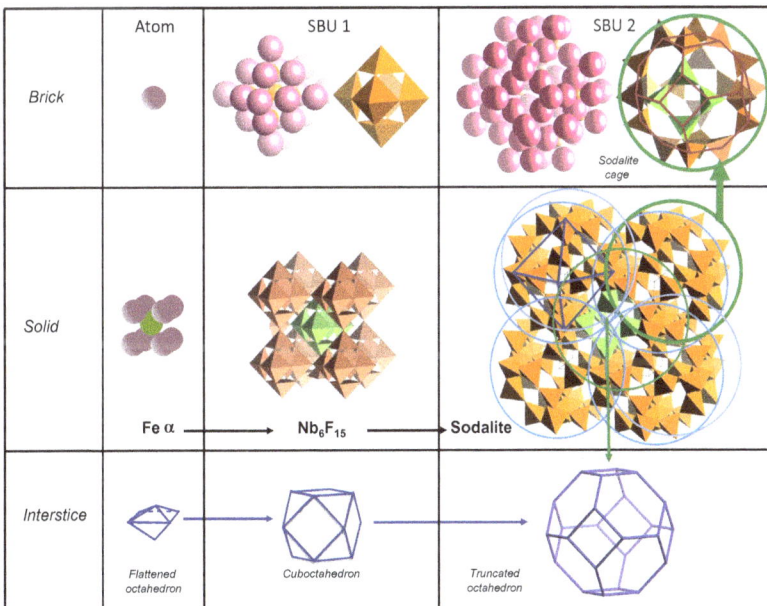

Fig. 7.2. Scale chemistry in I lattices. The green color in sodalite in the middle row is just here to highlight a I lattice.

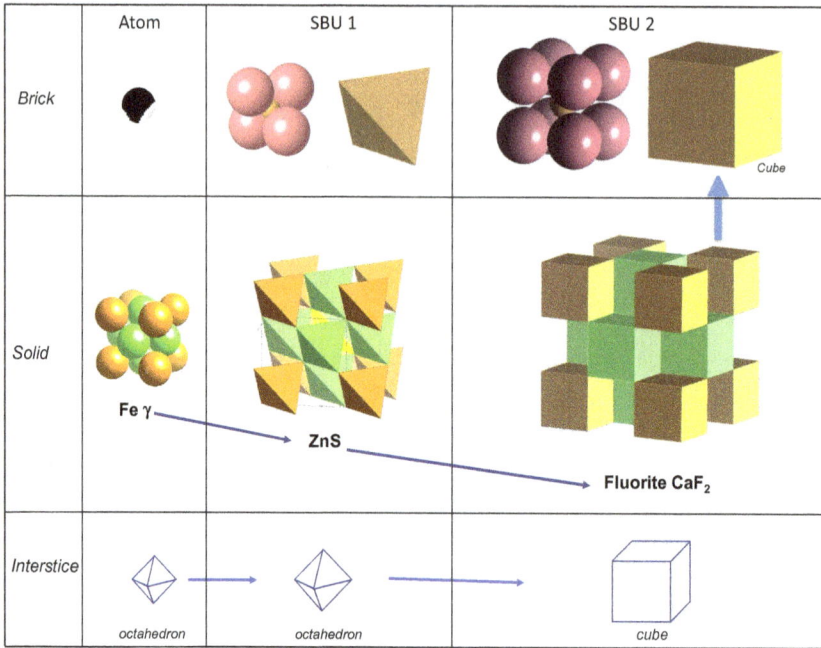

Fig. 7.3.   Scale chemistry in F lattices. The green color is just here to highlight a F lattice.

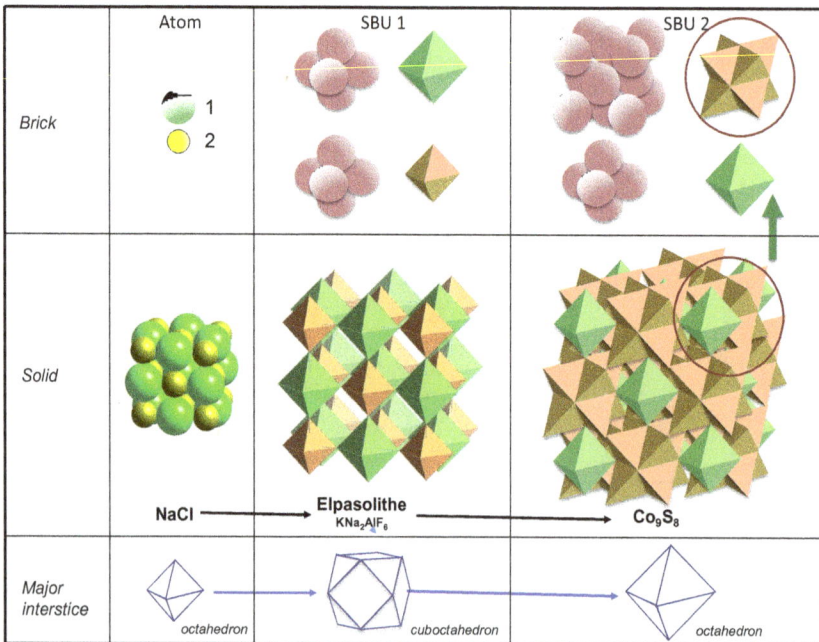

Fig. 7.4.   Scale chemistry in F lattices involving two ionic species, each in a *fcc* lattice.

These various examples begin to give a good idea of the possible applications of this concept. In other words, it is the connection of blocks, whatever their nature, which creates the structure by a periodic repetition. Within the type of connection chosen for these blocks, the chemical nature of the constitutive elements, the number of polyhedra (called *nuclearity* block), as their coordinations can be variable, but the topology of the arrangement remains invariant. This invariance is not restricted to cubic lattices; it applies to the seven crystalline networks (Fig. 7.5). Only the cell dimensions vary, at constant topology. This justifies the chosen term "scale chemistry" for qualifying this concept.

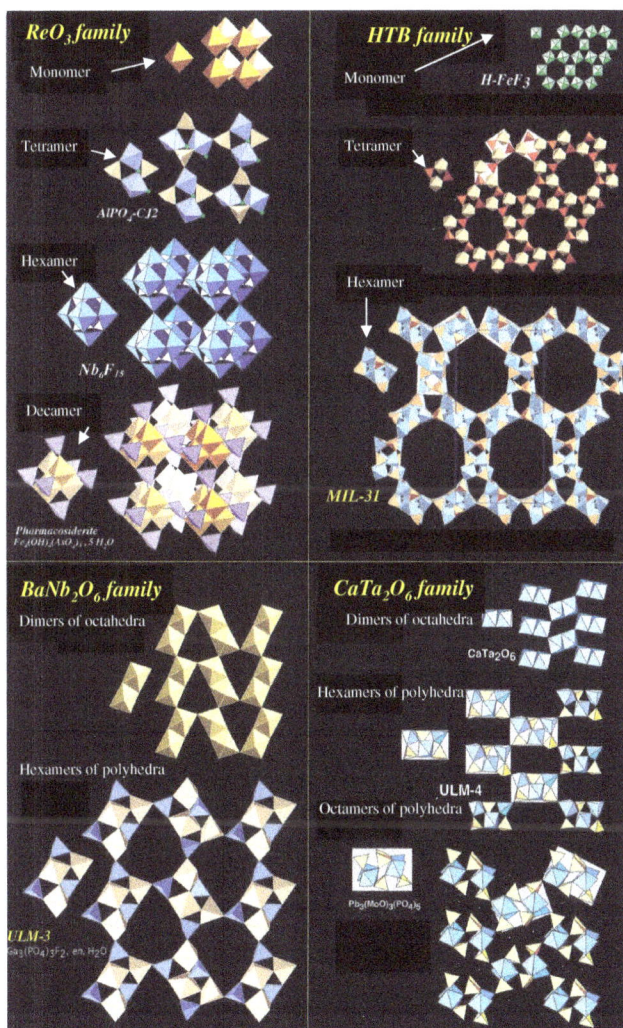

Fig. 7.5. Scale chemistry applied to various types of crystalline lattices in structural families already described in the book. Whatever the nuclearity of the SBU, the global topology remains invariant.

Once understood the principle and the extent of this concept, the chemist can now play on the SBUs and their constitution. Fixing the envelope, what can be placed inside, by translation, to create new solids? There are several ways of thinking over the problem.

7.1.1 *Filling the topological polyhedron*

The first way is to fill the initial topological polyhedron by different species. Take the example of the tetrahedron, starting brick for numerous solids, such as zinc sulfide ZnS or numerous silicates existing in nature. It can be filled first by other smaller tetrahedra, in such a way that the extremities of the bricks always allow the three-dimensional connection (Fig. 7.6).

In the series of indium selenides (Li, 1999a,b; Yaghi, 1994; Feng *et al.*, 2001, 2002; Férey, 2003), the super-tetrahedra, labelled T_n (n for the number of tetrahedra on the edge of the super-tetrahedron), can be filled or present vacancies [Fig. 7.7(a)]. A structure can present several types of super-tetrahedra [Fig. 7.7(b)]. The adopted structural type will depend on the nature of the counter-ions in the new structure.

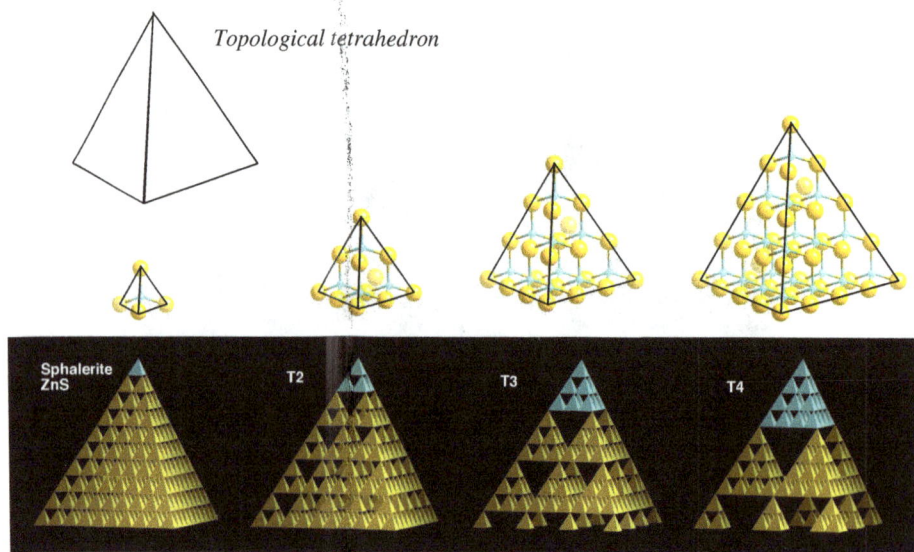

Topological tetrahedron

Fig. 7.6.   Scale chemistry applied to a topological tetrahedron. It can be filled by numerous smaller corner-shared tetrahedra, leading to a cubic sphalerite-type ZnS topology. This situation is encountered in a series of indium selenides.

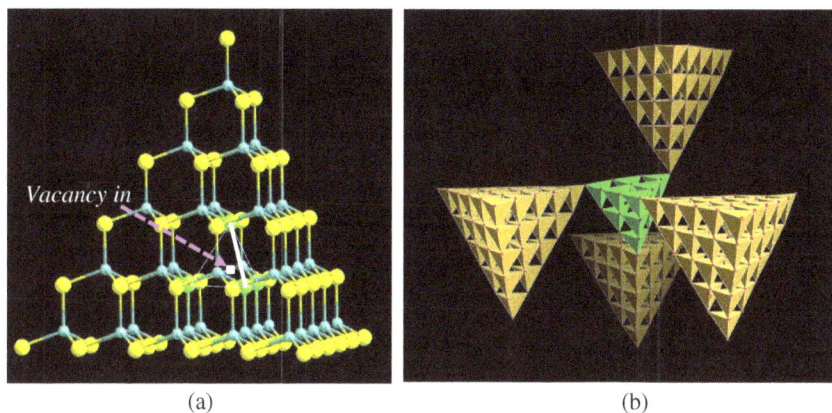

Fig. 7.7. (a) Creation of cationic vacancies in the tetrahedron; (b) Illustration of the possibility to connect super-tetrahedra of different nuclearities.

7.1.2 *The hybrid building blocks*

Instead of filling the topological tetrahedron, the second way will put SBUs at its vertices and link them by connectors, most of the time organic (Fig. 7.8). In terms of synthesis, this rather recent strategy presents many advantages. It indeed allows to expand the chemical possibilities for the constitution of the bricks (primitively purely inorganic), to a richer chemistry combining its inorganic and organic parts. This approach, born almost 15 years ago and called *hybrid chemistry*, offers quasi-infinite possibilities.

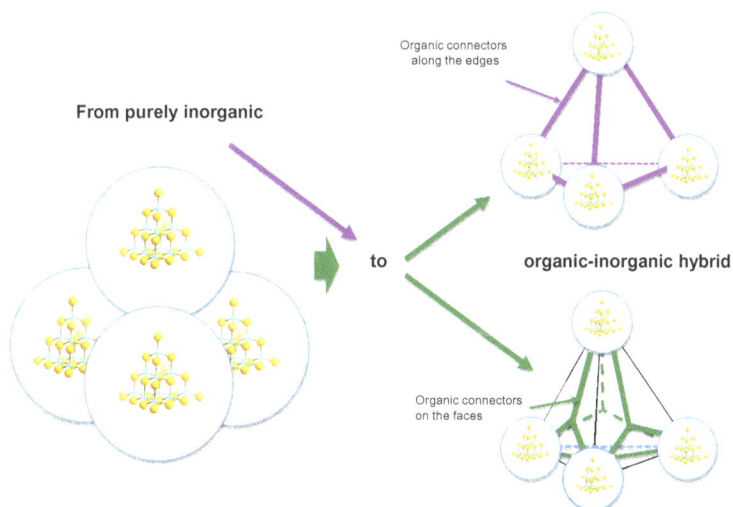

Fig. 7.8. From purely inorganic bricks to their hybrid equivalents; the organic connectors can be located either on the edges or on the faces of the topological tetrahedron.

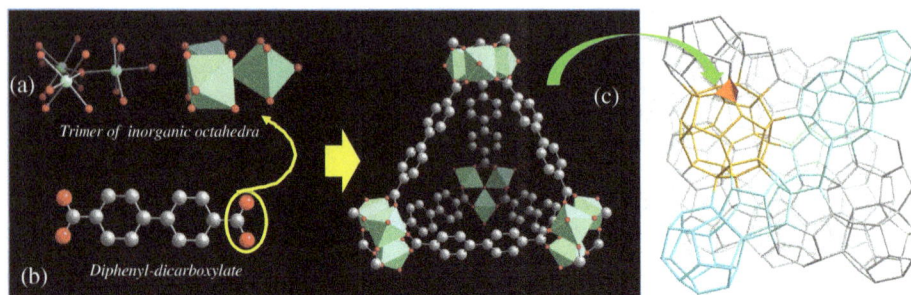

Fig. 7.9. Elaboration of the topological tetrahedron from the connection of a trimer of corner-sharing octahedra (a) with the anion diphenyl-dicarboxylate (b). The linkage indicated by the yellow arrow generates a giant tetrahedron (its edge is close to 16 Å) with the organic anion on the edges. The linkage by vertices of these super-tetrahedra leads to the complex three-dimensional structure MIL-101, created by the author. The figure on the right represents the skeleton of this structure, the lines symbolizing the connection of super-tetrahedra. Two types of cages limited to 20 (blue) and 28 (orange) super-tetrahedra are generated by this connection.

The organic connectors are usually polycarboxylates, polyphosphonates and nitrogen-based ligands. Figure 7.9 provides an example of the ability to link inorganic and organic constituents.

This topology was already known for one of the numerous varieties of SiO_2 (MTN type; MTN for Mobil Thirty-Nine). However, replacing a SiO_4 tetrahedron by the super-tetrahedron of Fig. 7.9(c) (its size is larger than that of the famous fullerene C_{60}) leads, keeping the topology, to a cell 1,000 times larger for the corresponding MIL-101 (MIL stands for Materials of Institut Lavoisier) than that of the silicate! This is scale-up chemistry (Férey, 2000, 2004, 2005)!

The introduction of organic linkers in these topologies has another advantage: the possibility to tune the length of the carbon chains between the complex terminal functions. This means that, within a given topology, the volume of the topological brick can rationally be increased or decreased, playing on the length and/or the hindrance of the linker.

Come back to the architecture of MIL-101 and its 28 tetrahedral members cage (Fig. 7.10). Starting from benzene 1,4-dicarboxylate, the increase of the length of the carbon chain up to diphenyl-dicarboxylate through naphtalene-dicarboxylate, leads to triple the volume of this cage, which reaches $60.000 Å^3$, a volume which could host 200 fullerenes C_{60}! In this solid, matter occupies only 8% of the space, with a density of 0.2.

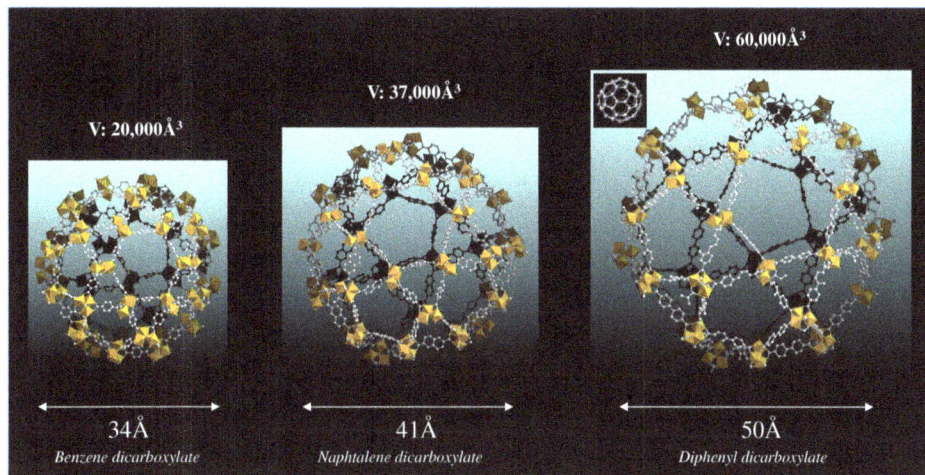

Fig. 7.10. Effect of the length of the carbon chain of the linker on the volume of the cages created by the corner-sharing of the hybrid super-tetrahedra. In each figure, the triangles correspond to the basis of the super-tetrahedra. In the figure on the right, the inset represents the fullerene C_{60} at the same scale.

7.1.3 *The anionic substitution in the inorganic bricks*

Scale chemistry can also be applied to the initial inorganic brick. The polyhedra inside it share an anion, often oxides, halides and sulfides, which are mono-atomic species. When these polyhedra share vertices, the two-fold coordination of the anions corresponds to M–O–M angles generally in the range 100° to 110°. A possible strategy is to substitute these anions by larger organic entities whose complex part also exhibits a two-fold coordination with respect to the cations. This increases the distance between the metallic cations and therefore the cell volume.

One organic entity is particularly efficient in this approach: the imidazolates, a vast family whose complex part is the function N–C–N, in which the nitrogen atoms provide the bonding with cations (Fig. 7.11). It was pioneered by Zhang and Chen, and Liu *et al.* in 2006.

Take for instance two ZnO_4 tetrahedra sharing a vertex. The Zn–Zn distance is close to 3 Å. If, instead of the oxide anion, an imidazole group is inserted (Fig. 7.11), the Zn–Zn distance becomes close to 6 Å and, consequently, the cell volume increases. In the example of Fig. 7.11, the topology was of the diamond type, but the concept can be applied to other topologies, mainly those of zeolites.

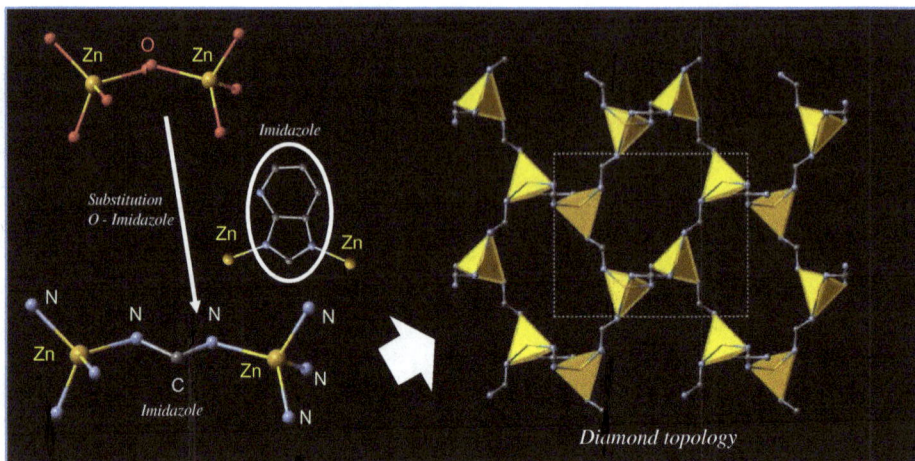

Fig. 7.11.   Scheme of the substitution of an oxide ion sharing ZnO_4 tetrahedra by the complex function N–C–N of imidazole, which maintains the diamond topology. For sake of clarity, the non-complex part of imidazole is not represented.

When the chosen topologies contain infinite chains of polyhedra instead of isolated ones, this substitution leads to a slightly different situation, as shown in Fig. 7.12 where the concept is applied to HTB topology.

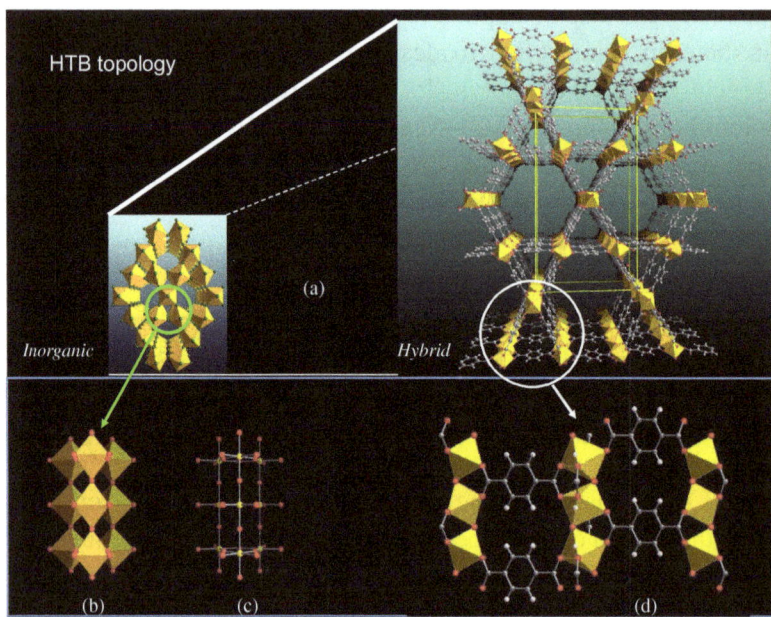

Fig. 7.12.   (a) Comparison between a purely inorganic HTB topology and its hybrid equivalent (the connector is benzene 1,4-dicarboxylate; (b,c) view of the trimeric chains within the green circle of (a) both in terms of octahedra (b) and of atoms and bonds (c); (d) the same for hybrid HTB.

Both the steric hindrance and the shape of the chelating function of carboxylate oblige on one hand the octahedral chains, primitively non-tilted, to become tilted; on the other hand, to a shift of the connectors [Fig. 7.12(d)].

7.1.4 *Decoration of polyhedra and construction of the inorganic brick*

Up to now, the inorganic bricks were considered as starting objects, but how are they formed, at least when they are polyatomic?

At least for some of them, it can be considered that they result from the **decoration** of Platonic or Archimedian polyhedra. Note that this notion of decoration (or augmentation), shared between the present approach (Férey and Cheetham), and that of M. O'Keeffe (Li *et al.*, 1999), in particular concerns two dimensions, the replacement of points in a geometric figure by surfaces, and in three dimensions, a vertex by a volume.

Take the example of the Archimedian truncated octahedron, seen in Chapter 1. It corresponds to one of the cases of truncation of the vertices of an octehedron, the other being the cuboctahedron. The decoration of the six square faces by tetrahedra sharing corners leads to a tetramer of tetrahedra and to the creation of the already described sodalite cage (Fig. 7.13). If one can imagine to replace the square face by another one consisting of four squares forming another square, its decoration by the same tetrahedra as above will lead, by scale chemistry, to a square of nine corner-sharing tetrahedra instead of four. The volume of the cage is now eight times the initial one, with always the same sodalite topology. The case exists for an aluminum zinc phosphate (MIL-74) in which Zn and Al exhibit a strict cationic order.

One can go further in this change of scale of the sodalite structure (Fig. 7.13 bottom row), as it was done for indium sulfides (Fig. 7.7), by considering each tetrahedron of the sodalite cage as a topological tetrahedron and its filling by an increasing number of "chemical" tetrahedra (here InS_4).

What happens for the cuboctahedron, the other Archimedian polyhedron resulting from the truncation of the octahedron? The decoration of its eight triangular faces by tetrahedra leads to an octamer of tetrahedra (already described as double four rings D4R), which exists in numerous zeolites. The eight terminal atoms (in black in Fig. 7.14) define a cube which will be the starting topological brick.

For inventing new structures, the imagination of the crystal chemist will play with the connections of the topological bricks, leading or not to original arrangements. For instance, starting from the above D4R, it is easy to *a priori* imagine a centered cubic arrangement for them (Fig. 7.15), with or without

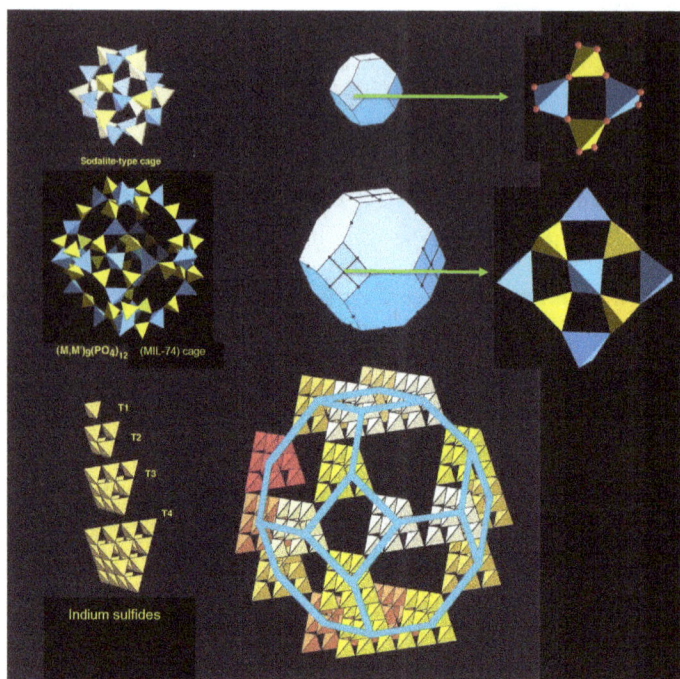

Fig. 7.13.   Scale chemistry applied to a truncated octahedron. It occurs in two steps, first by decoration of the six square faces of the truncated octahedron by corner-sharing tetrahedra, giving the sodalite topology (top row); the second step, after a doubling of the edge of the initial square faces, a decoration of the resulting four squares by the same tetrahedra as above. It leads to a volume of the resulting sodalite cage eight times larger than the first one. The down row presents an extension of the scale-up chemistry concept which, starting from the sodalite cage, replaces the single tetrahedra (considered here as the topologic tetrahedra) by increasing assemblies of corner-shared tetrahedra (noted by T_1 to T_4). The cage of the figure corresponds to T_3 encountered in Indium sulfides. Each topologic tetrahedron then contains ten initial tetrahedra.

a tilting between them. Experimentally, their organizations were found in cobalt phosphates.

For the moment, this approach is a game of imagination, strictly intellectual, as do children with Lego™, or architects, who, on the drawingboard in their office, draw sketches for representing a new building before choosing the technical solution taking into account other constraints. Now, with the progresses in informatics, these games can be modeled by simulating all the associations of topological bricks. These programs exist, but their details overcome the aims of this introduction to crystal chemistry. Nevertheless, Fig. 7.16 presents some of these results for cubes.

The use of such programs of simulation leads to thousands of possible solutions. Some of them are irrealistic for a crystal chemist; others seem more reasonable when compared to existing structures. Those which seem plausible according

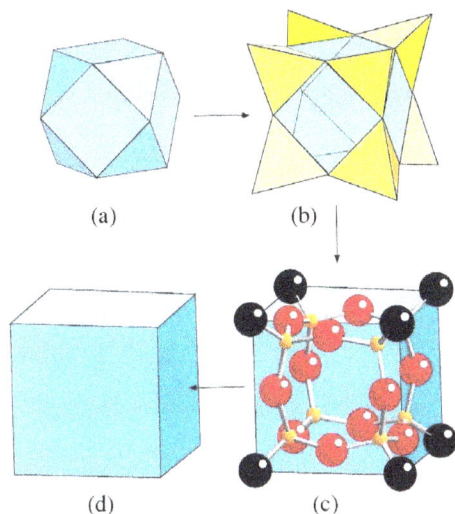

Fig. 7.14. (a) The cuboctahedron; (b) decoration of the triangular faces by regular tetrahedra; (c) Atoms and bonds representation of the eight tetrahedra (terminal atoms in black) which define a cube (d), which will act as the topological brick.

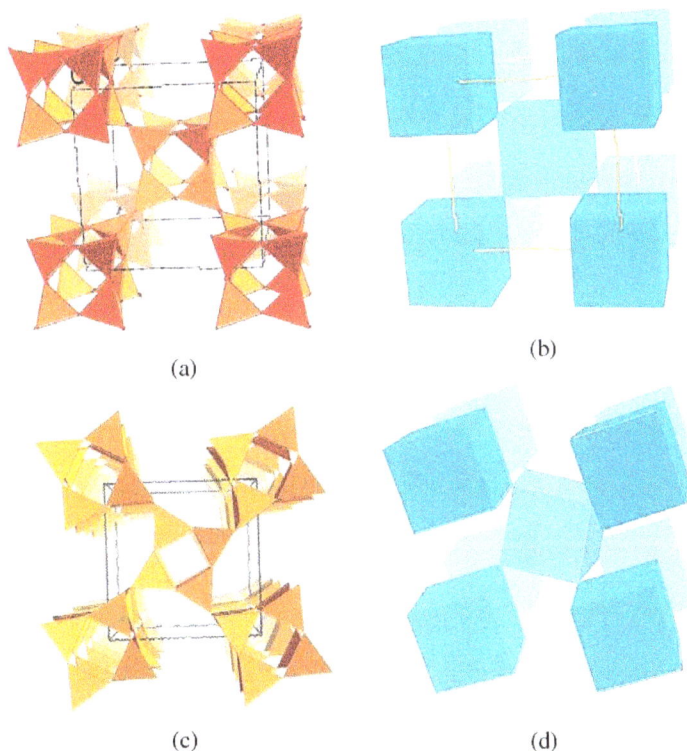

Fig. 7.15. Representation in terms of (a,c) D4R and of topologic cubes (b,d) of the non-tilted (top) and tilted (down) varieties of a cobalt phosphate. Phosphorus and cobalt, both in tetrahedral coordination, alternate in the D4R.

Fig. 7.16.   Some examples of virtual structures based on topological cubes corresponding to D4R units. The columns in orange represent these structures in terms of D4R and those in blue refer to the connection of topological cubes.

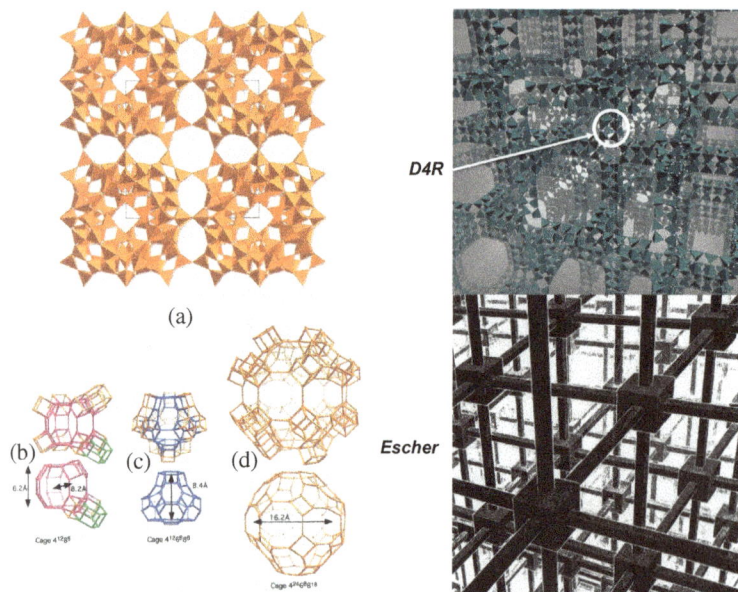

Fig. 7.17.   Two examples of virtual structures resulting from the association through vertices of topological cubes corresponding to D4R. In the structure of the left part (a) three kinds of cages (b,c,d) simultaneously appear, whereas already observed in other zeolites, as single examples. The structure on the right part recalls after authorization, a famous drawing of Escher (see text above).

to their lattice energy were called "*not-yet discovered structures*"... Figure 7.17 shows two examples. Owing to the classical knowledge on zeolites, the figure on the left seems reasonable. Despite its apparent complexity, one can distinguish, in the same structure, three kinds of cages, already found (but separately) in other zeolites. However, this structure would be the first to exhibit in the same architecture the three types of cages. In the upper right side of Fig. 7.17, another architecture appears. It is less likely but immediately reminds one of a drawing of Mauritius Escher, this Dutch artist at the beginning of the 20th century, famous for his strange drawings based on symmetry... Science and Art are clearly not incompatible, and the resonances between the two are numerous for those who know how to look!

7.2 The Concepts of Connectivity and Augmented Nets

As already said, this concept, due to Michael O'Keeffe (Li *et al.*, 1999), is close to that of scale chemistry. Their conclusions are identical. However, the elegance of his approach, more crystallographic, must be underlined. His starting point is the connectivity of the atoms of a parent structure.

Take his initial example, published in 1999: the structure of Pt_3O_4 (Fig. 7.18).

In it, platinum has a square planar coordination and oxygen atoms adopt a three-fold triangular one. O'Keeffe proposes to replace each atom of the structure by its connectivity figure: a square for platinum, a triangle for oxygen, and to connect the corners of these polygons by lines of connection (Fig. 7.19).

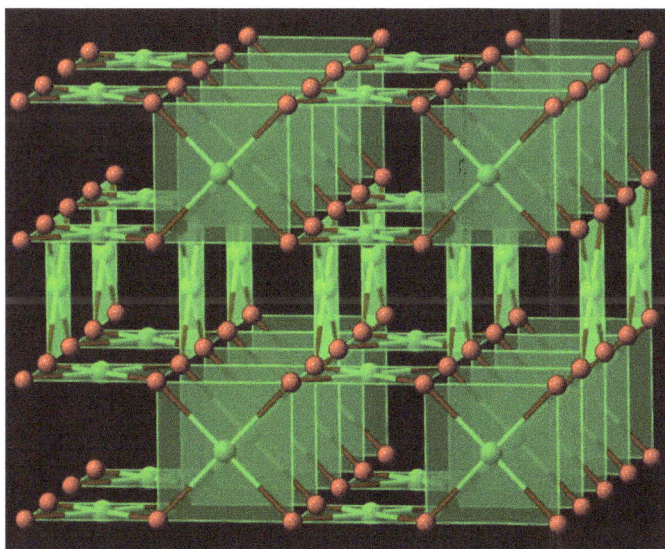

Fig. 7.18. Structure of Pt_3O_4 (Pt: green in a square coordination; O: red).

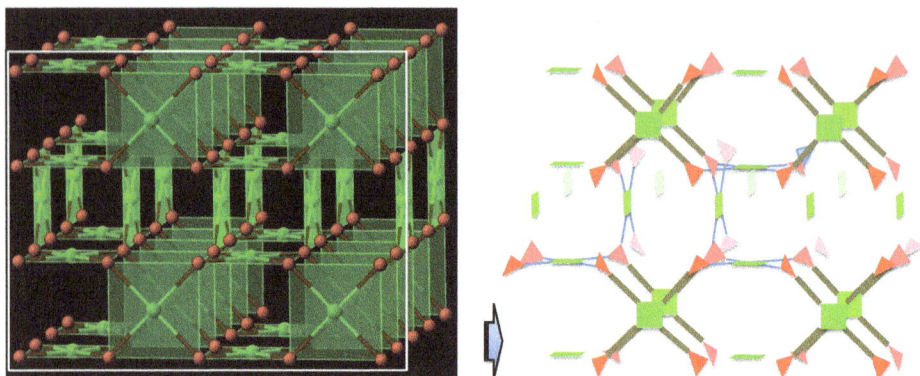

Fig. 7.19.   The O'Keeffe's vision of the structure of Pt_3O_4: (left) its classical polyhedral representation; (right) in terms of polygons of connectivity (Pt: green in a square coordination; O: red).

The game is now to find chemical species, whatever their dimensions, which present the same connectivity as the polygons of connectivity. Their association, which of course increases the dimensions of the objects compared to that of Pt_3O_4 gives rise to the so-called "**augmented nets**" (Fig. 7.20).

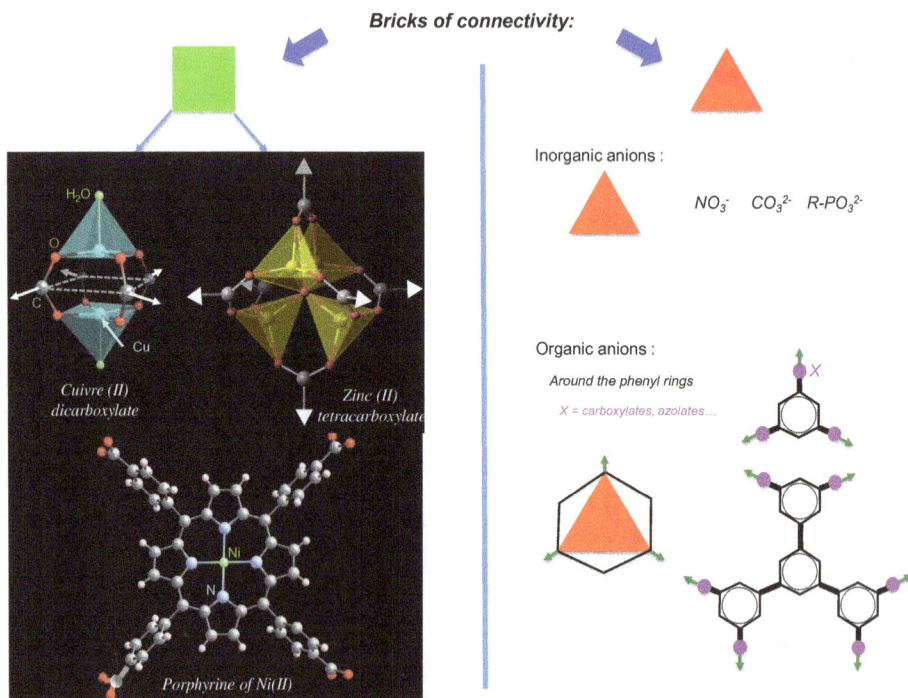

Fig. 7.20.   Some proposals of chemical species which have the same scheme of connectivity as the squares of the triangles of Fig. 7.19. With Zn, the brick can be either square or octahedral.

A nice example of association between square and triangular connectivity polygons is provided by HKUST-1 (Chui *et al.*, 1999; HKUST stands for Hong-Kong University of Science and Technology), a hydrated copper(II) benzene 1,3,5-tricarboxylate (hereafter noted by its acronym BTC) (Fig. 7.21). In this porous solid, the square polygon of connectivity corresponds to the four carbon atoms of the COO functions linking copper in the dimer of Fig. 7.21(a). In this dimer, copper exhibits a five-fold square pyramid coordination, due to the presence of a terminal water molecule bonded to Cu. When H_2O is eliminated by heating, the coordination of copper becomes square Fig. 7.21(d). The Cu–BTC linkages prove a three-dimensional porous structure with large cages [Figs. 7.21(b) and 7.21(c)].

This example, which concerns a three-fold connectivity for the linker and a square one for copper must not imply that they are alone to be discovered. Two-fold connectivity, often met for oxygen in oxides is also frequently encountered with

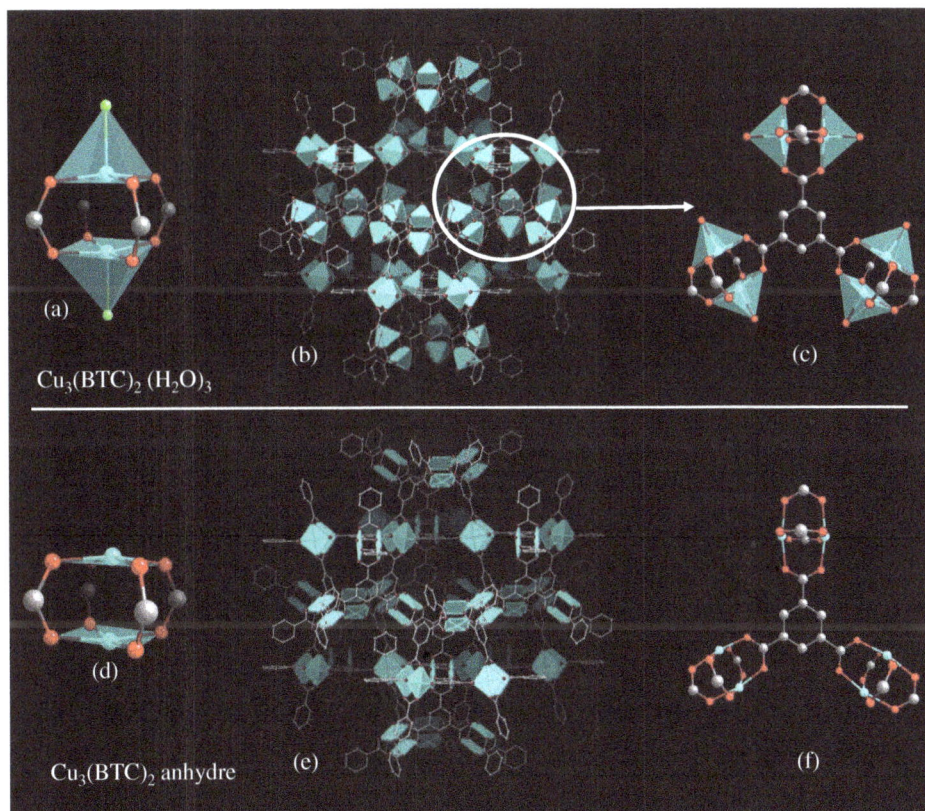

Fig. 7.21. In this copper(II) carboxylate, hydrated (b) or anhydrous (e), the chemical building block is a dimer of copper (a,d) whose aspect depends on the hydratation or not of the copper polyhedron; the connections between BTC and copper are identical in the two structures.

hybrid solids. When carboxylates are considered, this of course implies that they must have two opposite chelating functions, like for the terephtalate in Fig. 7.12.

The exchange oxygen–imidazolate in zeolite Rho is a good choice for illustrating the concept of augmented nets of O'Keeffe (Fig. 7.22).

Many structures of known simple solids can serve as parent structure types for the construction of augmented structures. The scale of homothety will essentially depend on the size of the organic part and on its steric hindrance. Indeed, other functions can be grafted on the organic molecule, while keeping the same distance between the complex functions.

It becomes clear that the two close concepts described above offer immense potential for the creation of a new chemistry, whatever its nature, inorganic, organic or hybrid. However, at this stage of the chapter, such potential remains entirely theoretical… For proving the actual efficiency of these concepts, the onset of experimental results is required and chemists face four main challenges:

— The application of these concepts would lead to structures with giant cells. Would they be thermally stable?
— What would be the interest of such giant cages?
— There are so many organic linkers that they do not represent a problem for the syntheses. It is not the case for inorganic bricks which are not readily usable

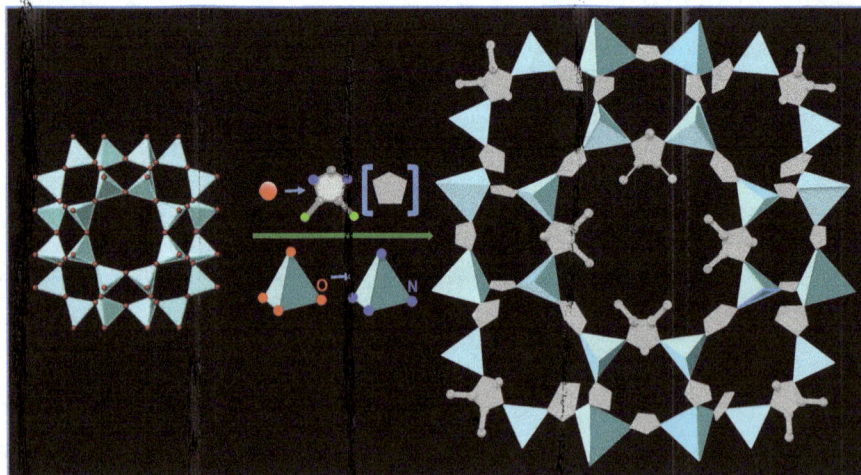

Fig. 7.22.   The structure of zeolite RhO (a hydrated sodium cesium aluminosilicate), augmented into its homologous zinc dichloro-imidazolate resulting from the substitution of oxygen by chloro-imidazolate. The SiO_4 tetrahedron becomes ZnN_4 because the nitrogens of the azole function are fixed on the metal (Zn). For the sake of clarity, the imidazolate is represented by a symbolic pentagon. By this substitution, the volume of the cubic cell becomes eight times that of the RhO structure.

as such: their shape and connectivity strongly depend on the conditions in which they are formed during the reaction. The question is therefore: how to find suitable inorganic building blocks?
— Once such solids are evidenced, how to transform these new chemical objects into materials?

In the 21st century, there is always Chemistry, but Materials Chemistry is also as a new objective...

7.3 From Concepts to Materials: The Different Steps

But, first, what is a material? It is at the origin a chemical compound, with a given structure, which exhibits various physical properties and interesting possible applications, but this definition is not sufficient. For becoming a material, several other conditions must be fulfilled. It must be stable versus temperature and humidity for practical use under several conditions. It must also be able to adopt different shapes (bulk, thin films, nanoparticles...) each adapted to a special property and use. Finally, for becoming a useful material, it must be cheap and easily prepared in large amounts, at the industrial scale, for its large diffusion in the society. The following paragraphs will provide some examples for the steps of this evolution.

A reply to the first challenge is positive. Despite established ideas, the hybrid solids with large pores experimentally obtained up to now have surprisingly a reasonable thermal stability in the range 25–400°C which allows a lot of applications.

The interest in giant cells? They clearly render the work of crystallographers more difficult but not impossible as soon as single crystals are obtained. Indeed, the fantastic technological progress (synchrotron radiation, high resolution electron microscopy, fast and efficient computer calculation programs) helps to solve these new structures and increase the interest of the chemist.

Some years ago, in rather dense inorganic structures, alkaline and alkaline-earth cations often occupied their interstices and ensured the electrical neutrality of the edifices. At constant topology, the important increase of the cell volumes in hybrid structures (and therefore of the interstices) renders the solids porous and makes other species, with a much larger volume and additional properties, to be incorporated in the pores and increase the range of chemical possibilities for creating new solids.

This implies a new vision for these solids. Besides the diverse occupations of the pores explained above, they exhibit also another feature, important in terms of materials: their specific surface area, at the interface between vacuum and matter. At variance to the atoms within the bulk, the atoms at this surface have not satisfied all their bonds with the surrounding neighbors. This creates an attractive potential facilitating the fixation of the molecules contained in the pores. Taking into account the time of residence of these molecules on the surface, this phenomenon will be used for example to separate quantitatively single molecules from an heterogenous mixture of them, with applications in chromatography. Moreover, as surface phenomena are important in catalytic reactions, the performances of these new solids must be tested in this domain (catalysis) which is of enormous economic importance.

7.3.1 *The inorganic building blocks*

A rational creation of new solids needs a good chemical and structural knowledge of these inorganic bricks and of their conditions of existence, function of the thermodynamic conditions applied to the synthesis. It is a very important question, difficult to solve, but less difficult if one takes a look at other disciplines.

Jean-Marie Lehn, the Nobel prize in Chemistry in 1987 said "*Il n'y a qu'une chimie!*" (there is only one chemistry). It seems evident, but so many chemists forget that! This is probably due to the fact that, with time, chemistry has exploded into various sub-disciplines: inorganic chemistry, organic chemistry, biochemistry, physical chemistry, coordination chemistry, supramolecular chemistry, materials chemistry... They all have experts, but this differentiation has led to such a compartmentalization that each category often ignores what happens within the others... It is time to reactivate interdisciplinarity and to restore a better chemical culture.

A major merit of the giant cells is to offer such opportunity for looking at other disciplines. In particular, the inorganic bricks are well known by coordination chemists from their molecular approach. Not only do they discover these bricks, characterize their morphology but also describe their connectivities and their conditions of synthesis. More than 130 inorganic bricks were recently reviewed (Tranchemontagne *et al.*, 2009), but only very few of them are currently used... This means that their eventual combination with the very large number of existing organic linkers leads to a quasi-infinity of possibilities for creating new porous solids. This justifies the enormous interest of the international community for this new and rich domain of chemistry.

7.3.2 *The pores and their occupation*

7.3.2.1 *General considerations*

This notion is intrinsic to porous solids. At the end of their syntheses (very often hydro- or solvothermal), the pores of the resulting solid are mainly filled by molecules of the solvent (H_2O or organic) used for the synthesis. The latter can be easily extracted by moderate heating (100–150°C). Once desolvated, the pores are empty and can be used for the introduction of various species (ionic or neutral). Whatever their nature, they bring other elements, and therefore other properties which did not exist initially. This ability to introduce post-synthesis new different species, is unique in the world of three-dimensional solids and at the origin of numerous current innovations.

For inserting species in pores, some conditions must be fulfilled. Besides the fact that the inserted species will not react with the starting solid, they both relate to the dimensions of the inside volume of the pore (largest as possible, playing on the length of the linker) and to their accessibility, determined by the size of the windows of access. The role of the windows is essential: (i) their size must be sufficiently large for allowing an easy passage of the guest molecules; (ii) when these guests correspond to a mixture of species, the windows of the pore must also be suffi-ciently controllable and tunable in diameter for serving as a molecular sieve for separating the entering species that could cause hindrance. They can (and also ought to) allow the further delivery of these species, except if they are used for capture.

The statement of Aristotle, *Nature abhors a vacuum* is a chance for the chemist! He has all the tools in his hands, temporarily or definitively. Figure 7.23 gives a first example of definitive insertion of polyanions in a porous structure (Armatas *et al.*, 2005). Once entered, the Mo_6O_{19} anion is trapped inside the phtalocyanine matrix because strong interactions exist with the latter. In the same way, the MIL-101 type (Fig. 7.24) (Férey, 2005) is able to insert five large Keggin ions per large pore of its structure.

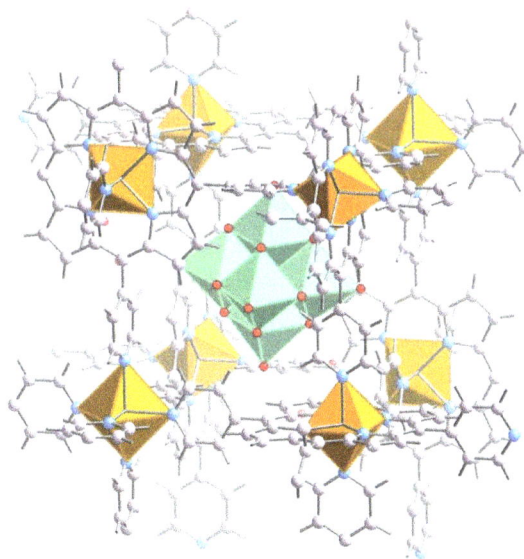

Fig. 7.23.   Insertion of a Mo_6O_{19} polyanion (in green) in an iron(III) phtalocyanine (Fe in orange).

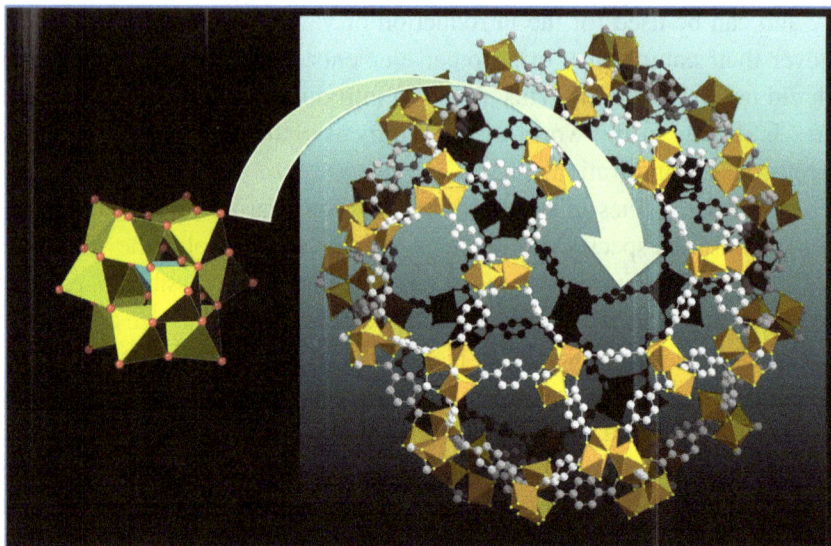

Fig. 7.24.   Insertion of the phosphomolybdate Keggin ion (left in yellow) in the chromium terephtalate MIL-101 (right). Five polyanions of this type can occupy one of the large cages of this solid.

The two above examples concern solid polycondensed anionic species, soluble in water. The anions enter the structure when the evacuated porous solid is dropped in a solution containing these anions. In the same way, the molecular solution can be poured on the solid which adsorbs the molecules of solvent. In every case, pores act as a trap, due to rather strong interactions with the pore wall. Depending on their energy, the introduction is reversible or not.

7.3.2.2 *Applications*

Indeed, the storage in the pore can be time-limited. All the time with no outside effect (temperature, vacuum, chemical exchange…) can lead to an evacuation, the host–guest assembly can be easily transported everywhere, while preventing the guest from any external attack, up to its delivery. The examples below relate to current applications in the domains of environment, energy and health.

When environment is concerned, the hybrid solids are excellent containers for storing greenhouse gases at room temperature, particularly CO_2. The metallic terephtalate MIL-101 (Fig. 7.24), is currently the best adsorber of this gas at room temperature. Introduced under some bars of pressure, very large amounts of CO_2 penetrate the structure (400 volumes of CO_2 per volume of product under 30 bars at 25°C) because chemists succeeded in making large pores. Not only are these amounts enormous, but they increase the storage capacities of conventional methods of storage. Among them, the storage in steel containers under high pressures of CO_2

is commonly used in industry. For instance, under a pressure of 30 bars, a steel container can store 50 volumes of CO_2 per volume unit but, if the container is filled by MIL-101 before introducing CO_2, 400 volumes of this gas can be stored per volume unit.... This means that, for a given volume of container, the introduction of this porous solid before any injection of CO_2 in this volume increases the amount of stored gas by close to one order of magnitude. This has now an incidence in economic terms for the transport of large amounts of CO_2.

The pores of the same solids are also used for energy and energy saving purposes. Here, two examples regarding these aspects:

— The storage of hydrogen, which can provide a substitute to petroleum for energetic needs, due to its exothermic enthalpy of oxidation. This gas is also one of the two gases extremely useful in fuel cells, the other being oxygen. However, on earth, hydrogen does not exist in the natural state. After producing it by the electrolysis of water, it is necessary to store it before eventual other uses, and porous solids with large pores can serve as its containers. The most promising was when they could already store 10% of their weight in hydrogen, which represents enormous volumes of H_2 gas. Unfortunately, such storage is only feasible at low temperature (77 K) and need expensive technological adaptations for its storage.

— Gasoline and fuel are by-products of petroleum; they are essential for vehicular traffic. The announced depletion of these resources obliges scientists to urgently find substitutes to this important energy product. In terms of circulation of cars, natural gas, whose main component is methane, is very abundant and can provide solutions. Indeed, its very high exothermicity during its oxidation allows the transformation of this thermal energy into an electrical one, applicable to cars. The only problem is to store natural gas in the car. Porous solids can efficiently play this role. This dream became a reality in 2013 thanks to BASF (Graab *et al.*, 2012). This company created a system based on an aluminum fumarate (isotypic with MIL-53) as container of natural gas. This led to a car with an autonomy of approximately 400km... Porous solids have become useful!

Finally, porous solids with giant pores find applications in a domain related to health: the nano-vectors for drugs. Nano-vectors are non-toxic nanometric species able to temporarily store drugs, transport them without damage to the body, reach the target (an ill organ) and deliver them efficiently.

The currently used nanovectors are liposomes (phospho-lipidic micelles). They store small amounts of drugs (ca. 5% in weight) within their small cages. Moreover, they are fragile in biological medium, and the delivery of drug is not quantitative. Instead of liposomes, the use of porous solids with large pores,

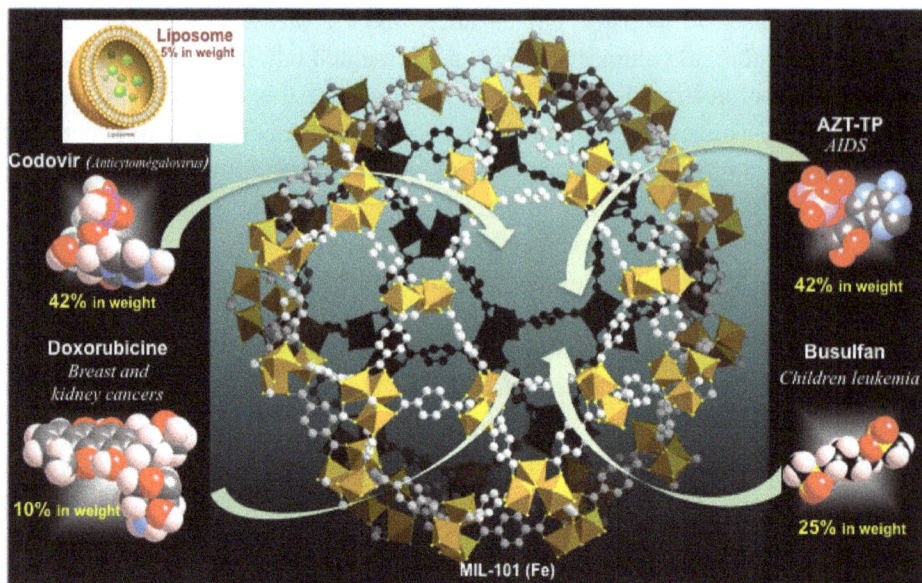

Fig. 7.25. Performances of the nanovector MIL-101 as carrier of drugs active against diverse types of cancer and AIDS, compared to those of the currently used liposomes.

flexible and rigid, was recently tested. Iron(III)-based MILs were chosen for their verified non-toxicity and their efficiency in medical imaging (Fig. 7.25). The tests were carried out on various drugs with different hydrophilicities, active against some types of cancers and AIDS. The results were promising because, for some drugs, the stored and (quantitatively) delivered amount of drugs is sometimes one order of magnitude larger than for liposomes... Thanks to scale-up chemistry, chemists were able to create giant pores, a trump for creative chemists!

Indeed, besides the properties of the skeleton of the frameworks, this existence of pores and their ability to insert different classes of species (ionic or neutral) adds new properties, compared to those of the initial porous compounds. These new solids provide a nice example of what is called now *multifunctional materials*. Once the filling of the pores is realized, the same solid cumulates several properties at the same time. For instance, MIL-101(Fe) is both a good catalyst, an excellent container for storage, and an interesting magnet...

7.3.3 *From the solid to the multi-material*

It was already stated that, for becoming a material, a solid, besides being a promising structure, must exhibit chemical and physical properties joined to an easy shaping, adapted to a dedicated function (thin films for sensors, nanoparticles…). This was thoroughly developed in the past for dense inorganic phases (mainly oxides), leading to a huge number of conductors (even superconductors), magnets and luminescent materials. A similar development for porous solids needs the creativity of chemists even for a given structure. But chemistry offers so many possibilities of substitution (anionic and cationic as well) that the challenge, favored by their adaptability at constant topology, will be successful. It just needs time and imagination.

A few examples can be illustrated. A porous solid, primitively insulating, can become a conductor under some conditions [Fig. 7.26 (left)], the first being that the metal in the skeleton must present multiple valences, what is the case for *3d* transition elements, like iron. If the starting material is based on a metal with a single oxidation state (e.g., Fe^{3+}), a classical electrochemical reduction by lithium metal (Férey *et al.*, 2007) generates a mixed valency (e.g. Fe^{3+}/Fe^{2+}), favoring electronic conduction by electron hopping between the two ions. Moreover, as the formed Li^+ cation has a small size, compared to that of the cages or the tunnels, it has important mobility. This adds an ionic conductivity to the electronic one, and the modified solid becomes a mixed conductor, because the cleverness of the chemist leads to a pertinent choice of the cation and the application of classical methods of electrochemistry…. Incidently, this experiment transforms the antiferromagnetic starting material into a ferrimagnetic one.

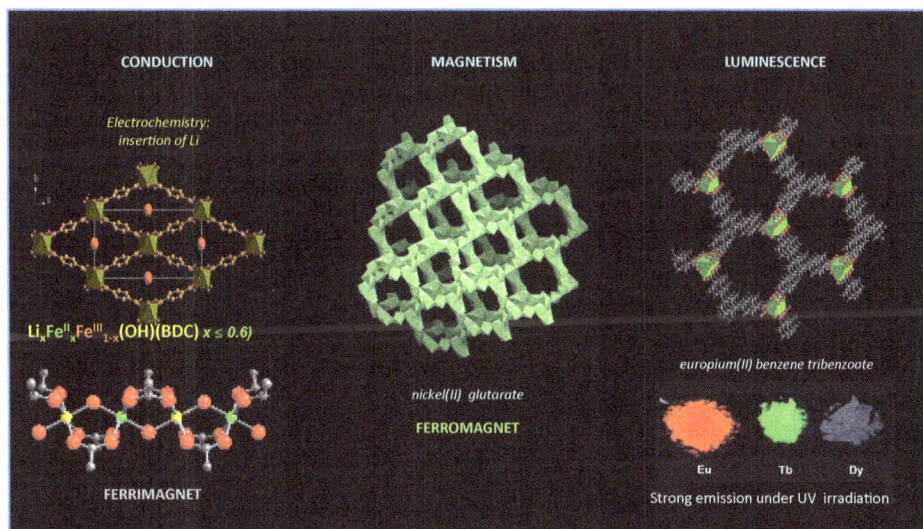

Fig. 7.26. Introduction of diverse physical properties in hybrid solids by an appropriate choice of cations or reducing elements.

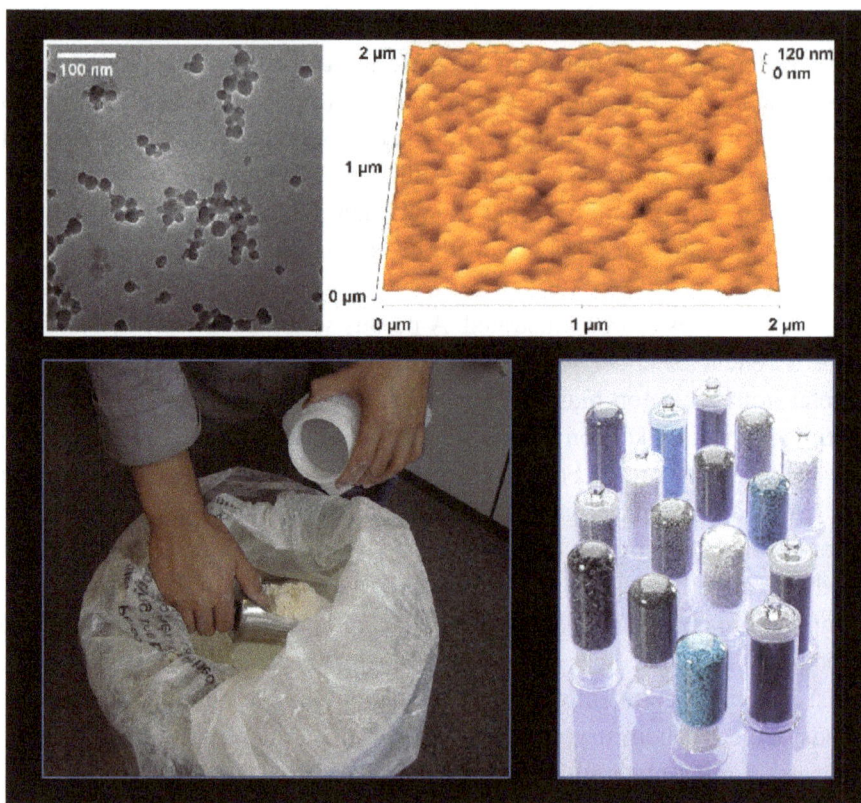

Fig. 7.27. Shaping of hybrid solids as nanoparticles, thin films, pills and their industrial production (here, 100 kg in one pot of aluminium terephtalate MIL-53(Al); with courtesy of BASF).

In the same way, magnetic properties (even ferromagnetic) can be generated by using *3d* transition metals and luminescent properties by using *4f* elements. This works, due to the talent and imagination of chemists and crystal chemists, opening new avenues for the future of chemistry and its applications.

Once the property is proved and/or conceptualized, dedicated shapings as a result of the aim of the property should be realized. Chemical engineering become necessary with, possibly, industrial cooperation for transforming these interesting academic solids into multifunctional materials useful for society (Fig. 7.27). When all the requirements are fulfilled, chemistry and crystal chemistry will become rational materials chemistry… The last example reveals that it is on its way!

Epilogue

The readers have now followed a long journey around the shapes. They began by a first contact with the simplest ones, those which, from Plato to Archimedes, they already knew. They discovered through their connections, their symmetries, their distortions, their defaults for establishing a bridge between the macroscopic and the molecular worlds. From a thorough analytical examination of the shapes, (which have become crystal structures), classifications were proposed. By the way, they also learnt the advantages to look at the structures in two ways: very close to their atomic organization, and from a distance, for understanding all the facets of the solid, facets which sometimes relate to art.

But crystal chemistry is not only analytic and descriptive. In the last chapter, it is shown that, using the understanding of these aspects, crystal chemistry is a powerful tool for a rational creation of new solids (the first concern of chemists!) and also for its applications for the benefit of the society.

This book aims to inspire young generations of scientists to be innovative, more rational and useful in chemistry. The author sincerely hopes that he has succeeded…

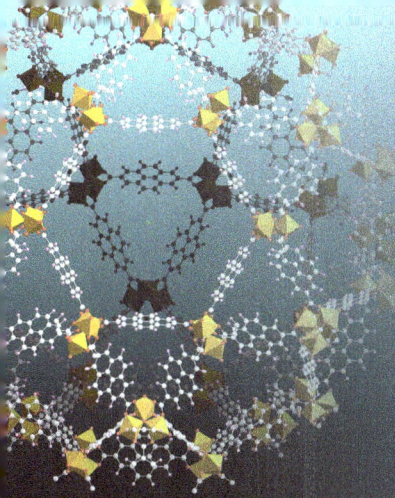

The Geometrical Characteristics of Platonic and Archimedian Polyhedra

This appendix provides all the geometric data relative to Platonic and Archimedian polyhedra and their dual forms (one page per polyhedron). In particular, the reader will find in this appendix all those relative to the tetrahedron and the octahedron, two polyhedra which were or will be frequently encountered in the different chapters. This avoids to repeat those boring calculations.

Each page contains, for each polyhedron:

- Its name and its Schläfli notation (see Chapter 1 for its definition).
- The number and the type of its faces, vertices and edges, the latter being noted A_{n-m}. n and m, equal or different, correspond to the type of faces which have this edge in common.
- Its dual form and dihedral angle(s) AD_{n-m}:
- Its angle(s) α between the two extremities of an edge and the center of the polyhedron.
- A/ρ: ratio between the length A of an edge and (ρ) of the distance between the center of the polyhedron and the middle of the edge.
- Length of the edge (normalized to 1). In dense packings, it corresponds to $2R$ (R: radius of the sphere).
- Distance between the center of the polyhedron and the center of its face.
- Distance between the center of the polyhedron and the middle of the edge.
- Distance between the center of the polyhedron and the vertex.
- Surface of the polyhedron = Σ surfaces of the faces, expressed in squares of edge length units. If the length of the edge is in Å, the surface is expressed in Å^2.
- Volume of the polyhedron: If the length of the edge is in Å, the volume is expressed in Å^3.

The perspective views and the planar developments of the dual forms of the Archimedian polyhedra are extracted from the book of Robert Williams: *The geometrical foundation of natural structure, a source book of design* (1972, 1979), Dover Editions, New York, NY 10014.

Platonic Solids

The tetrahedron: *Schläfli notation*: **[3,3]**

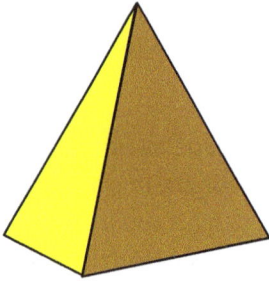

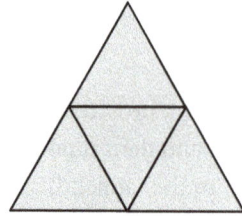

Faces: 4 (triangles)   Vertices: 4   Edges 3-3: 6

Dual form: tetrahedron   Dihedral angle AD$_{3\text{-}3}$: 70°31′44″ (arcsin($\sqrt{3}$)/3)

- Angle α (between the extremities of an edge and the center of the polyhedron): 109°28′16″ (arccos($-\frac{1}{3}$)).
- A/ρ: (length A of the edge/length ρ from the center of the polyhedron to the middle of the edge): $2\sqrt{2} = 2.828$.
- Normalized length of the edge: 1 (or 2R).
- Normalized distance from the center of the polyhedron to the center of the face: $\sqrt{6}/12 = 0.2041$.
- Normalized distance from the center of the polyhedron to the middle of the edge: $\sqrt{2}/4 = 0.3536$.
- Normalized distance from the center of the polyhedron to the vertex: $\sqrt{6}/4 = 0.6124$.
- Normalized distance from the center of a face to the vertex: $\sqrt{3}/3 = 0.5774$.
- Normalized distance from the center of a face to the middle of an edge: $\sqrt{3}/6 = 0.2887$.
- Surface of the polyhedron $\sqrt{3} = 1.732$.
- Volume of the polyhedron: $\sqrt{2}/12 = 0.1178$.

The way for obtaining these values was detailed in Chapter 1, during the calculation of the *r/R* ratio in the tetrahedral interstice of a face centered cubic (fcc) structure.

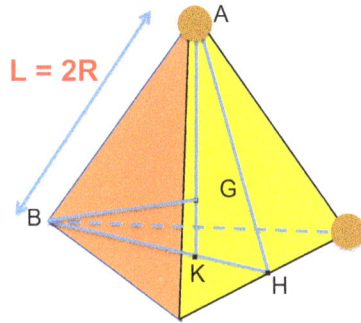

Platonic Solids

The cube: *Schläfli notation:* **[4,3]**

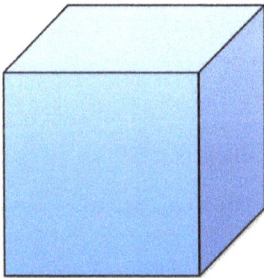

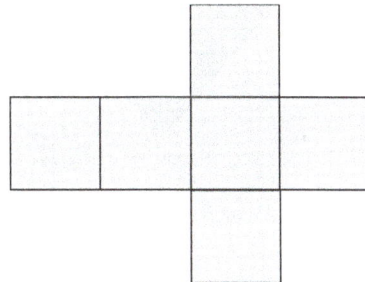

Faces: 6 (squares)   Vertices: 8   Edges$_{3-3}$: 12

Dual form: octahedron   Dihedral angle AD$_{4-4}$: 90°

- Angle α (between the extremities of an edge and the center of the polyhedron): 70°31′44″(arccos ($\frac{1}{3}$)).
- A/ρ: (length **A** of the edge/length ρ from the center of the polyhedron to the middle of the edge): $\sqrt{2} = 1.414$.
- Normalized length of the edge: 1 (or 2R).
- Normalized distance from the center of the polyhedron to the center of the face: ½.

- Normalized distance from the center of the polyhedron to the middle of the edge: $\sqrt{2}/2 = 0.7071$.
- Normalized distance from the center of the polyhedron to the vertex: $\sqrt{3}/2 = 0.8660$.
- Normalized distance from the center of a face to the vertex: $\sqrt{2}/2 = 0.7071$.
- Normalized distance from the center of a face to the middle of an edge: $\frac{1}{2} = 0.5$.
- Surface of the polyhedron 6.
- Volume of the polyhedron: 1.

Platonic Solids

The octahedron: *Schläfli notation:* [3,4]

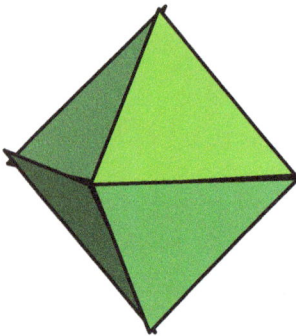

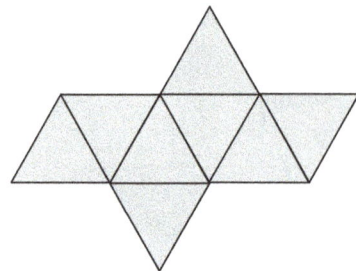

Faces: 8 (triangles)   Vertices: 6   Edges$_{3-3}$: 12

Dual form: octahedron   Dihedral angle AD$_{3-3}$: 109°28′16″ (arccos(−⅓))

- Angle α (between the extremities of an edge and the center of the polyhedron): 90°.
- A/ρ: (length A of the edge/length ρ from the center of the polyhedron to the middle of the edge): 2.
- Normalized length of the edge: 1 (or 2R).
- Normalized distance from the center of the polyhedron to the center of the face: $\sqrt{6}/6 = 0.4083$.
- Normalized distance from the center of the polyhedron to the middle of the edge: $\frac{1}{2} = 0.5$.
- Normalized distance from the center of the polyhedron to the vertex: $\sqrt{2}/2 = 0.7071$
- Normalized distance from the center of a face to the vertex: $\sqrt{3}/3 = 0.5774$.

- Normalized distance from the center of a face to the middle of an edge: $\sqrt{3}/6 = 0.2887$.
- Surface of the polyhedron $2\sqrt{3} = 3.4641$.
- Volume of the polyhedron: $\sqrt{2}/3 = 0.4714$.

Platonic Solids

The dodecahedron: *Schläfli notation*: $[5,3]$

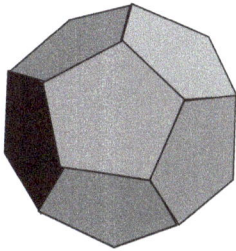

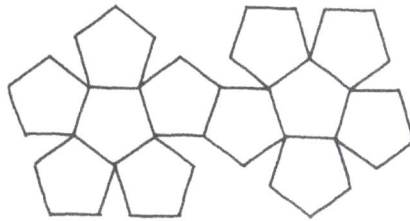

Faces: 12 (pentagonal)   Vertices: 20   Edges $_{5\text{-}5}$: 30

Dual form: icosahedron   Dihedral angle $AD_{5\text{-}5}$: 116°33′54″

- Angle α (between the extremities of an edge and the center of the polyhedron): 41°28′.
- **A/ρ:** (length **A** of the edge/length ρ from the center to the middle of the edge): $3 - \sqrt{5} = 0.7639$.
- Normalized length of the edge: 1 (or 2R).
- Normalized distance from the center of the polyhedron to the center of the face: $\tau^{5/2}/2(5^{1/4}) = 1.1135$ (τ is the gold number equal to $(1 + \sqrt{5})/2) = 1.6180$.
- Normalized distance from the center of the polyhedron to the middle of the edge: $\tau^2/2 = 1.309$.
- Normalized distance from the center of the polyhedron to the vertex: $\tau\sqrt{3}/2 = 1.4013$.
- Normalized distance from the center of a face to the vertex: $\sqrt{\tau}/(5^{1/4}) = 0.8507$.
- Normalized distance from the center of a face to the middle of an edge: $\sqrt{(25+10\sqrt{5})}/10 = 0.6882$.
- Surface of the polyhedron: $3\sqrt{(25+10\sqrt{5})} = 20.6458$.
- Volume of the polyhedron: $(15 + 7\sqrt{5})/4 = 7.6632$.

Platonic Solids

The icosahedron: *Schläfli notation:* **[3,5]**

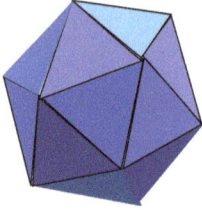

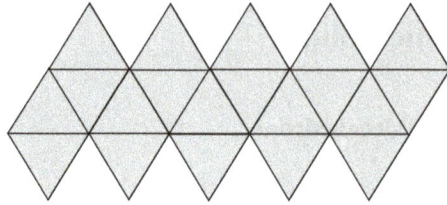

Faces: 20 (triangles)   Vertices: 12   Edges$_{3\text{-}3}$: 30

Dual form: dodecahedron   Dihedral angle AD$_{3\text{-}3}$: 138°11′22″

- Angle α (between the extremities of an edge and the center of the polyhedron): 63°26′ (arccos($\sqrt{5}$)/ 5).
- **A/ρ:** (length **A** of the edge/length ρ from the polyhedron center to the middle of the edge): $\sqrt{5}-1 = 1.2361$.
- Normalized length of the edge: 1 (or 2R).
- Normalized distance from the center of the polyhedron to the center of the face: $\tau^2/2\sqrt{3} = 0.7558$ (τ is the gold number: $(1 + \sqrt{5})/2 = 1.6180$).
- Normalized distance from the center of the polyhedron to the middle of the edge: $\tau/2 = 0.8090$.
- Normalized distance from the center of the polyhedron to the vertex: $\sqrt{3}/3 = 0.5774$.
- Normalized distance from the center of a face to the vertex: $\sqrt{3}/3 = 0.5774$.
- Normalized distance from the center of a face to the middle of an edge: $\sqrt{3}/6 = 0.2887$.
- Surface of the polyhedron $5\sqrt{3} = 8.6603$.
- Volume of the polyhedron: $5\tau^2/6 = 2.1816$.

Archimedian Solids

The truncated tetrahedron:

Notation de Schläfli: **[3.6²]**

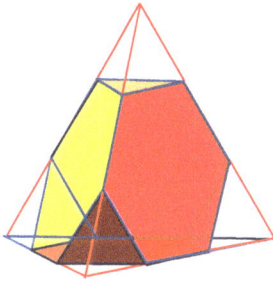

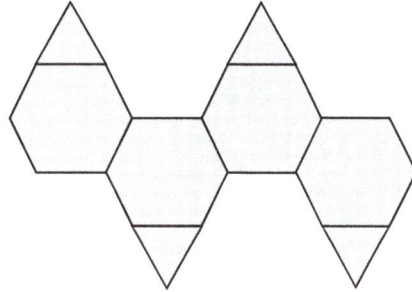

Faces: 4 (triangles)	Vertices: 12	Edges$_{6\text{-}6}$: 6
Faces: 4 (hexagons)		Edges$_{6\text{-}3}$: 12

Dual form: triakistetrahedron

Dihedral angle AD$_{3\text{-}6}$: 109°28′16″
Dihedral angle AD$_{6\text{-}6}$: 70°31′44″

- Angle α (between the extremities of an edge and the center of the polyhedron): 50°28′.
- **A/ρ:** (length **A** of the edge/length ρ from the polyhedron center to the middle of the edge): $\sqrt{8}\,/\,3 = 0.9428$.
- Normalized length of the edge: 1 (or 2R).
- Normalized distance from the center of the polyhedron to the center of the face: $\sqrt{6}\,/\,4 = 0.6124$.
- Normalized distance from the center of the polyhedron to the middle of the edge: $3\sqrt{2}\,/\,4 = 1.0607$.
- Normalized distance from the center of the polyhedron to the vertex: $\sqrt{22}\,/\,4 = 1.1726$.
- Normalized distance from the center of a triangular face to the vertex: $\sqrt{3}\,/\,3 = 0.5774$.
- Normalized distance from the center of a hexagonal face to the vertex: 1.
- Normalized distance from the center of a triangular face to the middle of an edge: $\sqrt{3}\,/\,6 = 0.2887$.
- Normalized distance from the center of a hexagonal face to the middle of an edge: $\sqrt{3}\,/\,2 = 0.8660$.
- Surface of the polyhedron 12.124.
- Volume of the polyhedron: 2.7102.

Archimedian Solids

The cuboctahedron:

Schläfli notation: $[3.4]^2$

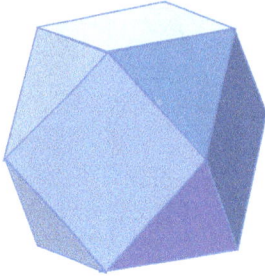

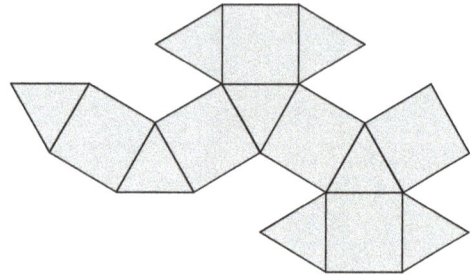

Faces: 8 (triangles)	Vertices: 12	Edges$_{3-4}$: 12
Faces: 6 (squares)		

Dual form: Rhombic dodecahedron   Dihedral angle AD$_{3-4}$: 125°15′51″ (arccos($-\sqrt{3}$/3))

- Angle α (between the extremities of an edge and the center of the polyhedron): 60°.
- A/ρ: (length **A** of the edge/length ρ from the polyhedron center to the middle of the edge) $2\sqrt{3}/3 = 1.1547$.
- Normalized length of the edge: 1 (or 2R).
- Normalized distance from the polyhedron center to the center of the triangular face: $\sqrt{6}/4 = 0.6124$.
- Normalized distance from the polyhedron center to the center of the square face: $\sqrt{2}/2 = 0.7071$.
- Normalized distance from the center of the polyhedron to the middle of the edge: $3\sqrt{2}/4 = 1.0607$.
- Normalized distance from the center of the polyhedron to the vertex: $\sqrt{3}/2 = 0.866$.
- Normalized distance from the center of a triangular face to the vertex: $\sqrt{3}/3 = 0.5774$.
- Normalized distance from the center of a square face to the vertex: $\sqrt{2}/2 = 0.7071$.
- Normalized distance from the center of a triangular face to the middle of an edge: $\sqrt{3}/6 = 0.2887$.
- Normalized distance from the center of a square face to the middle of an edge: ½ = 0.5.
- Surface of the polyhedron: 9.4641.
- Volume of the polyhedron: 2.3570.

View and development of the dual form: the non-Platonic rhombic dodecahedron (identical faces but different angles).

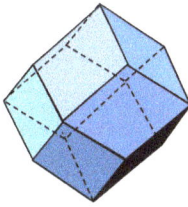

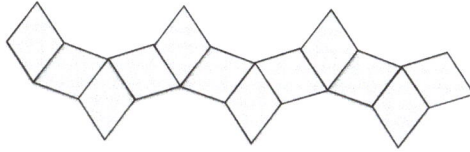

Archimedian Solids

The truncated cube: *Schläfli notation:* $\mathbf{[3.8^2]}$

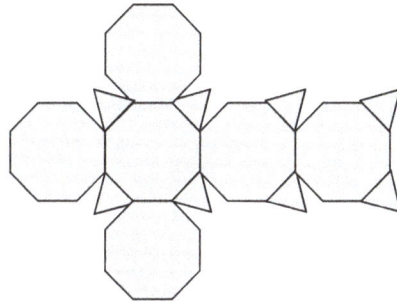

Faces: 8 (triangles)	Vertices: 24	$\text{Edges}_{3\text{-}8}$: 24
Faces: 6 (octogons)		$\text{Edges}_{8\text{-}8}$: 12

Dual form: triakisoctahedron

Dihedral angle: $\text{AD}_{3\text{-}8}$: $125°15'51''$ $(\arccos(-\sqrt{3})/3)$
Dihedral angle: $\text{AD}_{8\text{-}8}$: $90°$

- Angle α (between the extremities of an edge and the center of the polyhedron): $32°39$.
- A/ρ: (length **A** of the edge/length ρ from the polyhedron center to the middle of the edge) $2 - \sqrt{2} = 0.5858$.
- Normalized length of the edge: 1 (or 2R).
- Normalized distance polyhedron center — center of the triangular face: $(\sqrt{6}/12) \times (3\sqrt{2} + 4) = 1.6825$.
- Normalized distance polyhedron center — center of the octogonal face: $(1 + \sqrt{2})/2 = 1.2071$.
- Normalized distance from the center of the polyhedron to the middle of the edge: $(2 + \sqrt{2})/2 = 1.7071$.

- Normalized distance from the center of the polyhedron to the vertex: $\sqrt{(7/4)+\sqrt{2}} = 1.7787$.
- Normalized distance from the center of a triangular face to the vertex: $\sqrt{3}/3 = 0.5774$.
- Normalized distance from the center of an octogonal face to the vertex: $\frac{1}{2}\sqrt{\left(\sqrt{4+2\sqrt{2}}\right)} = 1.3066$.
- Normalized distance from the center of a triangular face to the middle of an edge: $\sqrt{3}/6 = 0.2887$.
- Normalized distance from the center of an octogonal face to the middle of an edge: $(1+\sqrt{2})/2 = 1.2071$.
- Surface of the polyhedron: 32.432.
- Volume of the polyhedron: 13.5988.

View and development of the (non-archimedian) dual form: the triakisoctahedron.

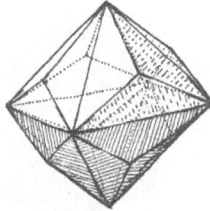

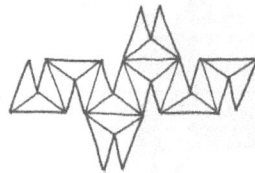

Archimedian Solids

The truncated octahedron: *Schläfli notation:* **4.6^2**

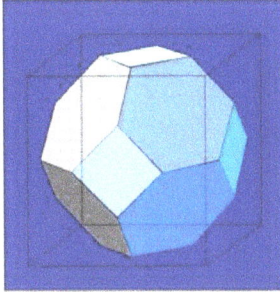

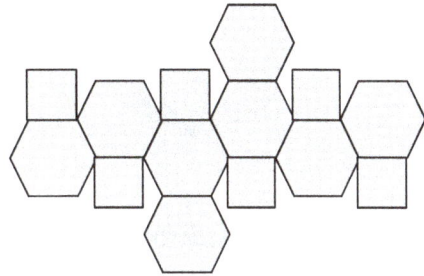

Faces: 6 (squares)	Vertices: 24	Edges$_{4\text{-}6}$: 24
Faces: 8 (hexagonal)		Edges$_{6\text{-}6}$: 12

Dual form: Tetrakis hexahedron   Dihedral angle AD$_{4\text{-}6}$: 125°15′51″ (arccos (−√3)/3)

Dihedral angle AD$_{6\text{-}6}$: 109°28′16″ (arccos(−⅓))

- Angle α (between the extremities of an edge and the center of the polyhedron): 36°52′.
- **A/ρ:** (length **A** of the edge/length ρ from the polyhedron center to the middle of the edge) $^2/_3 = 0.6667$.
- Normalized length of the edge: 1 (or 2R).
- Normalized distance from the polyhedron center to the center of the hexagonal face: $\sqrt{6}/2 = 1.2247$.
- Normalized distance from the polyhedron center to the center of the square face: $\sqrt{2} = 1.4142$.
- Normalized distance from the center of the polyhedron to the middle of the edge: 1.5.
- Normalized distance from the center of the polyhedron to the vertex: $\sqrt{10}/2 = 1.5811$.
- Normalized distance from the center of a hexagonal face to the vertex: 1.
- Normalized distance from the center of a square face to the vertex: $\sqrt{2}/2 = 0.7071$.
- Normalized distance from the center of a hexagonal face to the middle of an edge: $\sqrt{3}/2 = 0.866$.
- Normalized distance from the center of a square face to the middle of an edge: ½ = 0.5.
- Surface of the polyhedron: 26.7846.
- Volume of the polyhedron: 11.3137.

View and development of the (non-archimedian) dual form: the tetrakis hexahedron.

Archimedian Solids

Small rhombicuboctahedron: *Schläfli notation:* **3.4³**

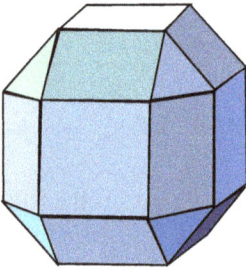

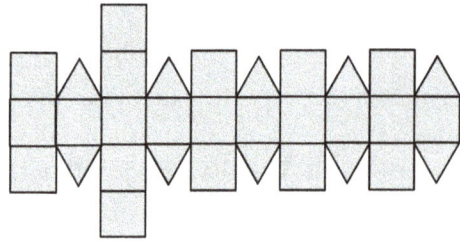

Faces: 8 (triangles)	Vertices: 24	Edges $_{3\text{-}4}$: 24
Faces: 18 (squares)		Edges $_{4\text{-}4}$: 24

Dual form: trapezoidal icositetrahedron   Dihedral angle AD$_{3\text{-}4}$: 144°44′08″(arccos(−$\sqrt{6}$)/3)
                                                 Dihedral angle AD$_{4\text{-}4}$: 135°(arccos(−$\sqrt{2}$)/2)

- Angle α (between the extremities of an edge and the center of the polyhedron): 41°53′.
- A/ρ: (length A of the edge/length ρ from the polyhedron center to the middle of the edge) $\sqrt{\left(2-\sqrt{2}\right)}$ = 0.7654.
- Normalized length of the edge: 1 (or 2R).
- Normalized distance polyhedron center — center of the triangular face: $\sqrt{6}[(3\sqrt{2}+2)/12]$ = 1.2743.
- Normalized distance polyhedron center — center of the square face: $(1+\sqrt{2})/2$ = 1.2071.

- Normalized distance from the center of the polyhedron to the middle of the edge $\sqrt{\left(1+\sqrt{2}/2\right)} = 1.3065$.
- Normalized distance from the center of the polyhedron to the vertex: $\sqrt{\left(5/4+\sqrt{2}/2\right)} = 1.3989$.
- Normalized distance from the center of a triangular face to the vertex: $\sqrt{3}/3 = 0.5774$.
- Normalized distance from the center of a square face to the vertex: $\sqrt{2}/2 = 0.7071$.
- Normalized distance from the center of a triangular face to the middle of an edge: $\sqrt{3}/6 = 0.2887$.
- Normalized distance from the center of a square face to the middle of an edge: $\frac{1}{2} = 0.5$.
- Surface of the polyhedron: 21.4641.
- Volume of the polyhedron: 8.7133.

View and development of the (non-archimedian) dual form: the trapezoidal icositetrahedron.

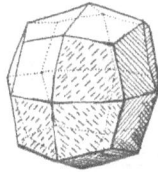

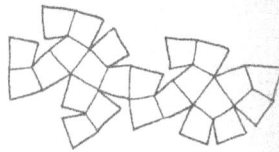

Archimedian Solids

Great rhombicuboctahedron: *Schläfli notation:* **3.6.8**

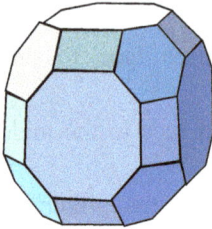   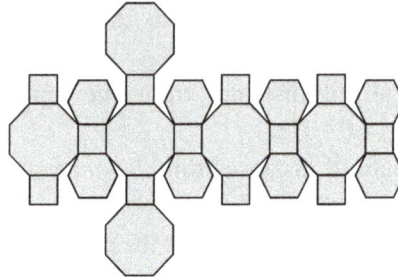

Faces: 12 (square)	Vertices: 48	Edges$_{4\text{-}6}$: 24
Faces: 8 (hexagonal)		Edges$_{4\text{-}8}$: 24
Faces: 6 (octogonal)		Edges$_{6\text{-}8}$: 24

Dual form: Hexakis octahedron

Dihedral angle AD$_{4\text{-}6}$: 144°44′08″ (arccos(−$\sqrt{6}$)/3)
Dihedral angle AD$_{4\text{-}8}$: 135° (arccos (−$\sqrt{2}$)/2)
Dihedral angle AD$_{6\text{-}8}$: 125°15′51″ (arccos(−$\sqrt{3}$)/3)

- Angle α (between the extremities of an edge and the center of the polyhedron): 24°55′.
- **A/ρ:** (length **A** of the edge/length ρ from the polyhedron center to the middle of the edge) $\sqrt{\left(2-\sqrt{2}\right)}/3 = 0.4419$.
- Normalized length of the edge: 1 (or 2R).
- Normalized distance from the polyhedron center to the center of the square face: $(3 + \sqrt{2})/2 = 2.2071$.
- Normalized distance polyhedron center — center of the hexagonal face: $\left[\sqrt{6}\left(\sqrt{2}+2\right)\right]/4 = 2.0908$.
- Normalized distance polyhedron center — center of the octogonal face: $(2\sqrt{2} + 1)]/2 = 1.9142$.
- Normalized distance from the center of the polyhedron to the middle of the edge: $3\sqrt{2}/4 = 1.0607$.
- Normalized distance from the center of the polyhedron to the vertex: $\left[\sqrt{\left(13+6\sqrt{2}\right)}\right]/2 = 2.3176$.
- Normalized distance from the center of a square face to the vertex: $\sqrt{2}/2 = 0.7071$.
- Normalized distance from the center of a hexagonal face to the vertex: 1.
- Normalized distance from the center of a octogonal face to the vertex: $\left[\sqrt{\left(4+2\sqrt{2}\right)}\right]/2 = 1.3066$.

- Normalized distance from the center of a square face to the middle of an edge: ½ = 0.5.
- Normalized distance from the center of an hexagonal face to the middle of an edge: $\sqrt{3}/2 = 0.866$.
- Normalized distance from the center of an octogonal face to the middle of an edge: $(1 + \sqrt{2})/2 = 1.2071$.
- Surface of the polyhedron: 61.7551.
- Volume of the polyhedron: 41.7942.

View and development of the (non-archimedian) dual form: the Hexakis octahedron.

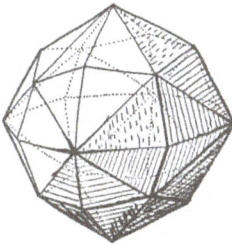

Archimedian Solids

Snub cube: *Schläfli notation:* $\mathbf{3^4.4}$*

  *According to the sense of rotation of the cube, two enantiomorphic forms (left and right) exist.

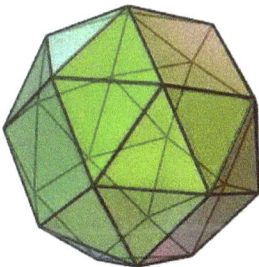

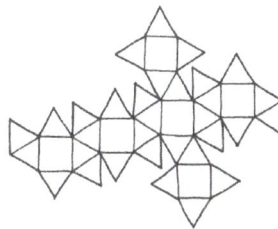

Faces: 32 (triangles)	Vertices: 24	Edges$_{3\text{-}3}$: 36
Faces: 6 (squares)		Edges$_{3\text{-}4}$: 24

Dual form: pentagonal icositetrahedron

Dihedral angle AD$_{3\text{-}3}$: 153°14′04″
Dihedral angle AD$_{3\text{-}4}$: 142°59

- Angle α (between the extremities of an edge and the center of the polyhedron): 43°41′.

- **A/ρ:** (length **A** of the edge/length ρ from the polyhedron center to the middle of the edge): 0.8018.
- Normalized length of the edge: 1 (or 2R).
- Normalized distance polyhedron center — center of the triangular face: 1.2132.
- Normalized distance from the polyhedron center to the center of the square face: 1.1425.
- Normalized distance polyhedron center — center of the triangular face: $\left[\sqrt{6} \left(\sqrt{2} + 2 \right) \right] / 4 = 1.1315$.
- Normalized distance from the center of the polyhedron to the middle of the edge: 1.2472.
- Normalized distance from the center of the polyhedron to the vertex: 1.3436.
- Normalized distance from the center of a triangular face to the vertex: $\sqrt{3} / 3 = 0.5774$.
- Normalized distance from the center of a square face to the vertex: $\sqrt{2} / 2 = 0.7071$.
- Normalized distance from the center of a triangular face to the middle of an edge: $\sqrt{3} / 6 = 0.2887$.
- Normalized distance from the center of an square face to the middle of an edge: ½ = 0.5.
- Surface of the polyhedron: 19.856.
- Volume of the polyhedron: 19.856.

View and development of the (non-archimedian) dual form: the right or left pentagonal icositetrahedron.

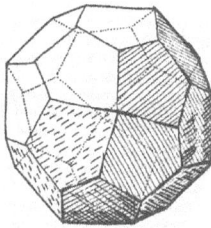

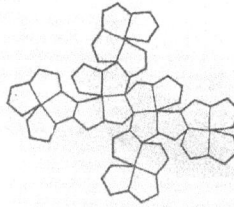

Archimedian Solids

The icosidodecahedron: *Schläfli notation*: $[3.5]^2$

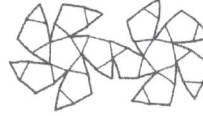

Faces: 20 (triangles)   Vertices: 30   Edges$_{3.5}$: 60
Faces: 12 (pentagonal)

Dual form: Rhombic triacontahedron   Dihedral angle AD$_{3-5}$: 142°37′21″

- Angle α (between the extremities of an edge and the center of the polyhedron): 36°.
- **A/ρ:** (length **A** of the edge/length ρ from the polyhedron center to the middle of the edge) $2\tan 18° = 0.6498$.
- Normalized length of the edge: 1 (or 2R).
- Normalized distance polyhedron center — center of the triangular face:
$$\sqrt{\left[\tau^2 - \left(1/\sqrt{3}\right)^2\right]} = 1.5083.$$
- Normalized distance polyhedron center — center of the pentagonal face:
$$\sqrt{\left[\tau^2 - 0.8806)^2\right]} = 1.3764.$$
- Normalized distance from the center of the polyhedron to the middle of the edge: 1.6180.
- Normalized distance from the center of the polyhedron to the vertex: $\tau = 1.6180$.
- Normalized distance from the center of a triangular face to the vertex: $\sqrt{3}/3 = 0.5774$.
- Normalized distance from the center of a pentagonal face to the vertex: $\sqrt{\tau}/\left(5^{1/4}\right) = 0.8507$.
- Normalized distance from the center of a triangular face to the middle of an edge: $\sqrt{3}/6 = 0.2887$.
- Normalized distance center of a pentagonal face to the middle of an edge: $\sqrt{(25 + 10\sqrt{5})}/10 = 0.6882$.
- Surface of the polyhedron: 29.3002.
- Volume of the polyhedron: 13.8237.

View and development of the (non-archimedian) dual form: the Rhombic triacontahedron.

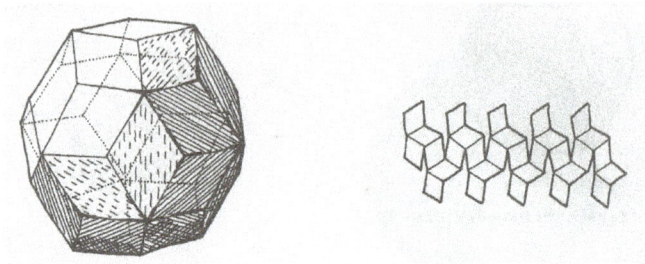

Archimedian Solids

The truncated dodecahedron: *Schläfli notation:* **3.10^2**

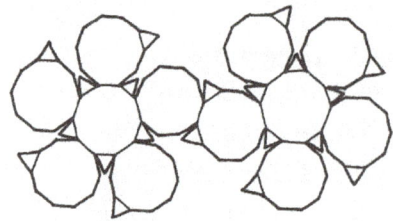

| Faces: 20 (triangles) | Vertices: 60 | Edges$_{3\text{-}10}$: 60 |
| Faces: 12 (decagons) | | Edges$_{10\text{-}10}$: 30 |

Dual form: Triakis icosahedron

Dihedral angle $AD_{3\text{-}10}$: 142°37′21″
Dihedral angle $AD_{10\text{-}10}$: 116°33′54″

- Angle α (between the extremities of an edge and the center of the polyhedron): 19°24.
- **A/ρ:** (length **A** of the edge/length ρ from the polyhedron center to the middle of the edge) $\left(3/\sqrt{5}\right) -1 = 0.3416$.
- Normalized length of the edge: 1 (or 2R).
- Normalized distance from the polyhedron center to the center of the triangular face: 2.9132.
- Normalized distance from the polyhedron center to the center of the decagonal face: 2.4904.

- Normalized distance from the center of the polyhedron to the middle of the edge: 2.9274.
- Normalized distance from the center of the polyhedron to the vertex: 2.9698.
- Normalized distance from the center of a triangular face to the vertex: $\sqrt{3}/3 = 0.5774$.
- Normalized distance from the center of a decagonal face to the vertex: $\tau = 1.6180$.
- Normalized distance from the center of a triangular face to the middle of an edge: $\sqrt{3}/6 = 0.2887$.
- Normalized distance center of a decagonal face — middle of an edge: $\left[\sqrt{\left(5+2\sqrt{5}\right)}\right]/2 = 1.5388$.
- Surface of the polyhedron: 100.9882.
- Volume of the polyhedron: 85.0542.

View and development of the (non-archimedian) dual form: the Triakis icosahedron.

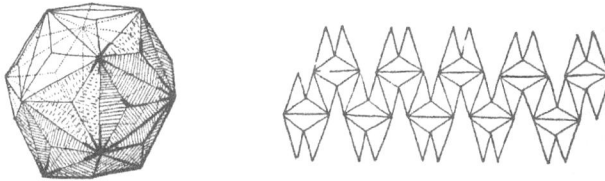

Archimedian Solids

The truncated icosahedron: *Schläfli notation:* **5.6^2**

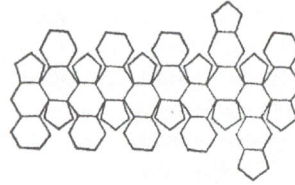

Faces: 12 (pentagons)	Vertices: 60	Edges$_{5\text{-}6}$: 60
Faces: 20 (hexagons)		Edges$_{6\text{-}6}$: 30

Dual form: Pentakis dodecahedron,   Dihedral angle AD$_{3\text{-}5}$: 142°37′21″
                                                  Dihedral angle AD$_{3\text{-}5}$: 138°11′22″

- Angle α (between the extremities of an edge and the center of the polyhedron): 23°17.
- A/ρ: (length A of the edge/length ρ from the polyhedron center to the middle of the edge) $\left(\sqrt{5}-1\right)/3 = 0.4120$.
- Normalized length of the edge: 1 (or 2R).
- Normalized distance from the polyhedron center to the center of the pentagonal face: 2.3276.
- Normalized distance from the polyhedron center to the center of the hexagonal face: $3\tau^2/2\sqrt{3} = 2.2672$.
- Normalized distance from the center of the polyhedron to the middle of the edge: 1.2472.
- Normalized distance from the center of the polyhedron to the vertex: 2.4782.
- Normalized distance from the center of a pentagonal face to the vertex: $\sqrt{\tau}/\left(5^{1/4}\right) = 0.8507$.
- Normalized distance from the center of a hexagonal face to the vertex: 1.
- Normalized distance center of a pentagonal face — middle of an edge: $\sqrt{(25+10\sqrt{5})}/10 = 0.6882$.
- Normalized distance from the center of a hexagonal face to the middle of an edge: $\sqrt{3}/2 = 0.8660$.
- Surface of the polyhedron: 72.6000.
- Volume of the polyhedron: 55.2870.

View and development of the (non-archimedian) dual form: the Pentakis dodecahedron.

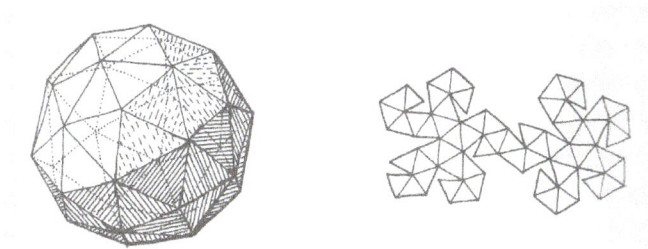

Archimedian Solids

The small rhombicosidodecahedron: *Schläfli notation*: **3.4.5.4**

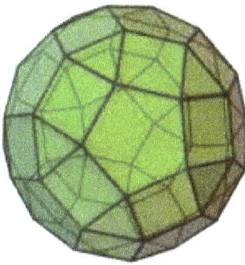

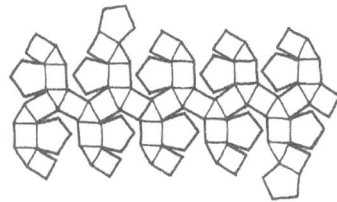

Faces: 20 (triangular)	Vetrices: 60	Edges $_{3-4}$: 60
Faces: 30 (square)		Edges $_{4-5}$: 60
Faces: 12 (pentagonal)		

Dual form: Trapezoidal hexacontahedron

Dihedral angle AD$_{3-4}$: 159°05′41″
Dihedral angle AD$_{4-5}$: 148°16′57″

- Angle α (between the extremities of an edge and the center of the polyhedron): 25°52′.
- A/ρ: (length **A** of the edge/length ρ from the polyhedron center to the middle of the edge) $\sqrt{2}$.tan 18° = 0.4595.
- Normalized length of the edge: 1 (or 2R).
- Normalized distance from the polyhedron center to the center of the triangular face: 2.1572.
- Normalized distance from the polyhedron center to the center of the square face: 2.1182.

- Normalized distance from the polyhedron center to the center of the pentagonal face: 2.0647.
- Normalized distance from the center of the polyhedron to the middle of the edge: 2.1763.
- Normalized distance from the center of the polyhedron to the vertex: 2.2351.
- Normalized distance from the center of a triangular face to the vertex: $\sqrt{3}/3 = 0.5774$.
- Normalized distance from the center of a square face to the vertex: $\sqrt{2}/2 = 0.7071$.
- Normalized distance from the center of a pentagonal face to the vertex: $\sqrt{\tau}/(5^{1/4}) = 0.8507$.
- Normalized distance from the center of a triangular face to the middle of an edge: $\sqrt{3}/6 = 0.2887$.
- Normalized distance from the center of a square face to the middle of an edge: $\frac{1}{2} = 0.5$.
- Normalized distance center of a pentagoanl face middle of an edge: $\sqrt{(25+10\sqrt{5})}/10 = 0.6882$.
- Surface of the polyhedron: 59.3002.
- Volume of the polyhedron: 41.6144.

View and development of the (non-archimedian) dual form: the trapezoidal hexacontahedron.

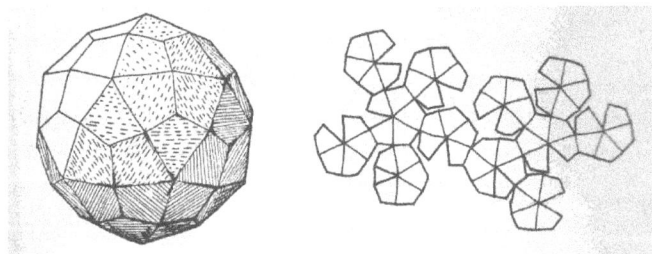

Archimedian Solids

Thee great rhombicosidodecahedron: *Schläfli notation*: **4.6.10**

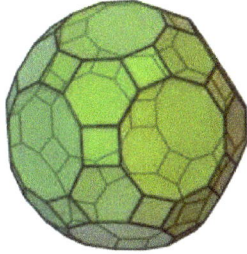

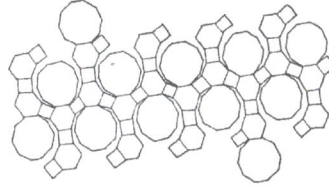

Faces: 30 (squares)   Vertices: 120   Edges$_{4-6}$: 60
Faces: 20 (hexagons)   Edges$_{4-10}$: 30
Faces: 12 (decagons)   Edges$_{6-6}$: 30

Dual form: Hexakis icosahedron   Dihedral angle AD$_{4-6}$: 159°05′41″
                                                Dihedral angle AD$_{4-10}$: 148°16′57″
                                                Dihedral angle AD$_{3-5}$: 142°37′21″

- Angle α (between the extremities of an edge and the center of the polyhedron): 15°6′.
- **A/ρ**: (length **A** of the edge/length ρ from the polyhedron center to the middle of the edge) $\sqrt{6}/3.\tan 18° = 0.2653$.
- Normalized length of the edge: 1 (or 2R).
- Normalized distance from the polyhedron center to the center of the square face: 3.7358.
- Normalized distance from the polyhedron center to the center of the hexagonal face: 3.6683.
- Normalized distance from the polyhedron center to the center of the decagonal face: 3.4407.
- Normalized distance from the center of the polyhedron to the middle of the edge: 3.7693.
- Normalized distance from the center of the polyhedron to the vertex: 3.8021.
- Normalized distance from the center of a square face to the vertex: $\sqrt{2}/2 = 0.7071$.
- Normalized distance from the center of a hexagonal face to the vertex: 1.
- Normalized distance from the center of a decagonal face to the vertex: $\tau = 1.6180$.
- Normalized distance from the center of a square face to the middle of an edge: ½.
- Normalized distance from the center of a hexagonal face to the middle of an edge: $\sqrt{3}/2 = 0.8660$.

- Normalized distance center of a decagonal face middle of an edge: $\left[\sqrt{\left(5+2\sqrt{5}\right)}\right]/2 = 1.5388$.
- Surface of the polyhedron: 174.2880.
- Volume of the polyhedron: 206.7839.

View and development of the (non-archimedian) dual form: the hexakis icosahedron.

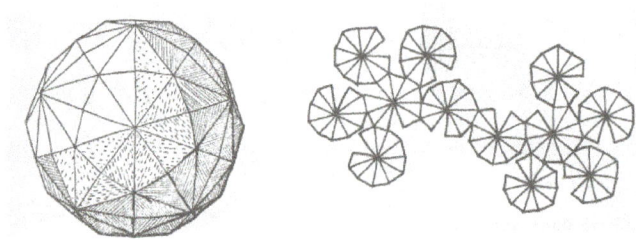

Archimedian Solids

The snub dodecahedron: *Schläfli notation:* $3^4.5$*

*According to the sense of rotation of the dodecahedron, two enantiomorphic forms (left and right) exist.

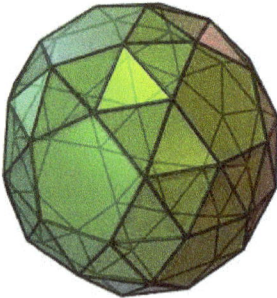
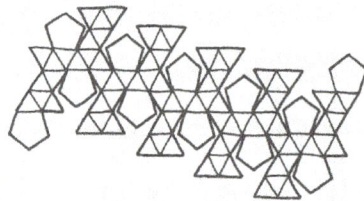

Faces: 80 (triangles)	Vertices: 60	Edges$_{3-3}$: 90
Faces: 12 (pentagons)		Edges$_{3-5}$: 60

Dual form: pentagonal hexacontahedron

Dihedral angle AD$_{3-3}$: 164°10′31″
Dihedral angle AD$_{3-5}$: 152°55′53″

- Angle α (between the extremities of an edge and the center of the polyhedron): 26°49′.
- A/ρ: (length **A** of the edge/length ρ from the polyhedron center to the middle of the edge): 0.4769.
- Normalized length of the edge: 1 (or 2R).

- Normalized distance from the polyhedron center to the center of the triangular face: 2.0768.
- Normalized distance from the polyhedron center to the center of the pentagonal face: 1.9806.
- Normalized distance from the center of the polyhedron to the middle of the edge: 2.0969.
- Normalized distance from the center of the polyhedron to the vertex: 2.1556.
- Normalized distance from the center of a triangular face to the vertex: $\sqrt{3}/3 = 0.5774$.
- Normalized distance from the center of a pentagonal face to the vertex: $\sqrt{\tau}/\left(5^{1/4}\right) = 0.8507$.
- Normalized distance from the center of a triangular face to the middle of an edge: $\sqrt{3}/6 = 0.2887$.
- Normalized distance center of a pentagonal face — middle of an edge: $\sqrt{(25+10\sqrt{5})}/10 = 0.6882$.
- Surface of the polyhedron: 55.2808.
- Volume of the polyhedron: 37.6072.

View and development of the (non-archimedian) dual form: the pentagonal hexacontahedron.

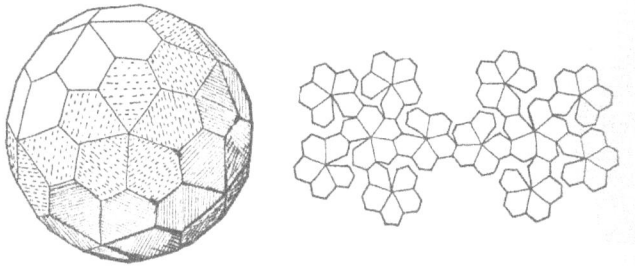

The Exponential Scale, or the Mathematical Nature of Polyhedra

The reader could be surprised to see mathematical developments in such a book, considered as an initiation to crystal chemistry. An initiation, it is like entering the hall of a house. An initiation, it is to give an idea of what will be in the house, without visiting all the rooms. At least, the destination of these rooms is written on the closed doors for information. The reader can stay only in the hall, but can also, if he is curious, open some of those closed doors to improve his knowledge…

This book is a hall; the current appendix opens one of these closed doors for just a glimpse for the reader. He can enter the "mathematical" door…. For that, as soon as he has assimilated the contents of this book, he can read the books and articles cited in the short (but essential) list of references…

Kant said that it was the Nature itself — and not the mathematicians — which led mathematics to natural philosophy…

May be! But mathematicians (and in a broader sense, scientists in general), continuously tried, over the centuries, to explain its secrets. It is always true, even in crystal chemistry. Indeed, it is worthy to note that, even if regular polyhedra are known and used from the ancient Greeks, no mathematical treatment of the analytical expression of these volumes is proposed. These polyhedra were only defined as starting objects on which the talent of scientists was applied for tracking down their properties, and never as mathematical functions of space variables.

At the end of the nineties, two Swedish chemists, Sten Andersson, and his young colleague Michael Jacob, realized this property. As indicated in the preface, Sten Andersson, was one the members of the quatuor who, in the seventies, had already philosophically understood the structural organization of the solid. Being a chemist and a crystallographer, he also had a strong mathematical background. He was passionate about inorganic structures, fascinated by their shapes and had become one of the rare virtuosi of the simplified description of these edifices. His vision of the solid was already considered by the community as exceptional, even visionary. Not him! He felt that it was necessary to go farther. He felt that these shapes, even the simplest, had to be justified mathematically by analytical functions using variables as the spatial coordinates in a reference trihedron. And he did it in both a series of articles published between 1994 and 1997, and in three superb books: *The language of the shape* (1997; Elsevier, Amsterdam); *The exponential scale* (1997; Oldenbourg Verlag, München) and *The nature of mathematics and mathematics of Nature* (1998; Elsevier, Amsterdam).

The starting principle is simple and begins with analytical geometry. This appendix aims only to understand his strategy, and not the mathematical developments.

In a reference trihedron, it is known that the functions $x = 0$ and $y = 0$, represent planes which intersect in space. With the function $10^x + 10^y = $ constant (called *isosurface constant*), one passes continously from one plane to the other. Then, by using power functions of 10 (10 or any other base, except 0 or 1) an additional rule is introduced, which implies different functions whose sum leads to another figure, this time predictible. The extension to the third dimension by adding 10^z defines the three basic planes of the reference trihedron. Without coming to the details, the reconstitution of the polyhedron defined by the three unit vectors needs to add the inverse functions at the primitive sum. For example, the function:

$$10^x + 10^y + 10^z + 10^{-x} + 10^{-y} + 10^{-z}$$

approximately represents a cube.

All the discovery of Andersson was to find the functions which mathematically define in the best way the Platonic and Archimedian (among others!). Instead of the 10 base, he chose the exponential base due to the multiplicity of the properties of this function. The results are remarkable, with increasing sophistication of the analytical expression of the functions.

Below, and just for information, appears the list of the best functions for some of the polyhedra encountered in this book.

Cube: $\exp(x^4) + \exp(y^4) + \exp(x^4) = 10^{10}$

Tetrahedron: $\exp\{(x + y + z)\}^3 + \exp\{(x - y - z)\}^3 + \exp\{(-x - y + z)\}^3 + \exp\{(-x + y - z)\}^3 = 4.\,10^4.$

Octahedron: $\exp\{(x + y + z)\}^4 + \exp\{(x - y - z)\}^4 + \exp\{(-x - y + z)\}^4 + \exp\{(-x + y - z)\}^4 = 4.\,10^3.$

Rhombic dodecahedron: $\exp\{(x + y)\}^6 + \exp\{(x - y)\}^6 + \exp\{(x + z)\}^6 + \exp\{(-x + z)\}^6 + \exp\{(z + y)\}^6 + \exp\{(-z + y)\}^6 = 10^4.$

Truncated octahedron: $\exp\{(x + y + z)\}^4 + \exp\{(x - y - z)\}^4 + \exp\{(-x - y + z)\}^4 + \exp\{(-x + y - z)\}^4 + \exp x^8 + \exp y^8 + \exp z^8 = 10^7.$

Truncated cube: $\exp\{(x + y + z)\}^2 + \exp\{(x - y - z)\}^2 + \exp\{(-x - y + z)\}^2 + \exp\{(-x + y - z)\}^2 + \exp x^8 + \exp y^8 + \exp z^8 = 10^5$.

Truncated tetrahedron: $\exp\{(x + y + z)\}^2 + \exp\{(x - y - z)\}^2 + \exp\{(-x - y + z)\}^2 + \exp\{(-x + y - z)\}^2 + \exp\{(x + y + z)\}^3 + \exp\{(x - y - z)\}^3 + \exp\{(-x - y + z)\}^3 + \exp\{(-x + y - z)\}^3 = 4.\ 10^6$.

References

- Articles (*the number of stars indicates the importance of each paper*).
1. **M. O'Keeffe, S.Andersson: "Rod packings and crystal chemistry" *Acta Crystallogr.* **1977**, *A33*, 914.
2. *S. Andersson, S.T. Hyde, K. Larsson, S. Lidin: "Minimal surfaces and structures: from inorganic and metal crystals to cell membranes and biopolymers" *Chem. Rev.* **1988**, *88*, 221–242.
3. ***B. Hyde, S. Andersson, "*Inorganic Crystal Structures*" (Wiley, New-York, 1988).
4. S. Lidin, M. Jacob, S. Andersson, "A mathematical analysis of rod packings" *J. Solid State Chem.* **1995**, *114*, 36.
5. ***S.T. Hyde, S. Andersson, K. Larsson, Z. Blum, T. Landh, S. Lidin, B. Ninham: *The Language of Shape*; the role of curvature in condended matter: physics, chemistry, and biology (Elsevier, Amsterdam, 1997).
6. **M. Jacob: "Saddle, tower and helicoidal surfaces" *J. Phys II* (France), **1997**, *7*, 1035–1044.

- Related to the exponential scale.
1. S. Andersson, M. Jacob, S. Lidin, "On the shape of crystals" *Z. Kristallogr.* **1995**, *210*, 3–4.
2. S. Andersson, M. Jacob, K. Larsson, S. Lidin: "Structure of the cubosome — a closed lipid bilayer aggregate" *Z. Kristallogr.* **1995**, *210*, 315–318.
3. S. Andersson, M. Jacob, S. Lidin: "The exponential scale and crystal structures" *Z. Kristallogr.* **1995**, *210*, 315–318.
4. K. Larsson, M. Jacob, S. Andersson: "Lipid bilayer aggregate standing waves in cell membranes" *Z. Kristallogr.* **1996**, *211*, 875–878.
5. M. Jacob, K. Larsson, S. Andersson: "Lipid bilayer standing waves conformations in aqueous cubic phases" *Z. Kristallogr.* **1997**, *212*, 5–8.
6. S. Andersson, M. Jacob: "On the structure of mathematics and crystals" *Z. Kristallogr.* **1997**, *212*, 334–346.
7. M. Jacob, S. Andersson: "Finite periodicity and crystal structures" *Z. Kristallogr.* **1997**, *212*, 486–492.
8. M. Jacob, S. Andersson: "Finite periodicity chemical systems and Ninham forces" *Colloids and Surfaces* **1997**, *129–130*, 227–237.

Bibliography

Useful books:

Burns G.; Glazer A.M., *Space groups for solid state scientists,* 2nd Ed. (Academic Press, New York, 1990).

Cambridge Crystallographic Data Centre, www.ccdc.cam.uk/.

Galasso F.S., *Structures and properties of inorganic solids*, (Pergamon Press, Oxford 1970).

Galasso F.S., *Perovskites and High T_c superconductors*, (Gordon & Breach Pubs., New York 1990).

Hyde B.G.; Andersson S., *Inorganic crystal structures*, (John Wiley & Sons, New York, 1989).

International tables for X-Ray crystallography, Volume A (Kluwer Academic, Dortrecht, 1992).

Mitchell R.H., *Perovskites, Modern and ancient* (Almar Press Inc., Thunder Bay, 2002).

O'Keeffe M.; Hyde B.G., *Crystal structures, I. Patterns and symmetry* (Mineral. Soc. of Amer., Washington, 1996).

Rao C.N.R.; Raveau B., *Transition metal oxides* (VCH Publishers, New York, 1995).

Raveau B.; Michel C.; Hervieu M.; Groult D., *Crystal chemistry of high-T_c superconducting copper oxides* (Springer-Verlag, Berlin, 1991).

Wells A.F., *Structural Chemistry*, 5th Ed. (Clarendon Press, Oxford, 1984).

Wyckoff R.W.G., *Crystal structures,* Vols. 1–4 (John Wiley & Sons, New York, 1968).

Some pioneering articles:

Armatas G.; Burkholder E.; Zubieta J., *J. Solid State Chem.* (2005) 178, 2430.

Bertaut E.F.; Blum P.; Sagnières A., *Acta Crystallogr.* (1959), 12, 149.

Birch W.D.; Pring A.; Reller A.; Schmalle H., *Naturwiss.* (1992), 79, 509.

Brese N.F.; O'Keeffe M., *Acta Crystallogr.* (1991), *B47*, 192.

Chui S.S.Y.; Lo S.M.F.; Charmant J.P.H.; Orpen A.G.; Williams I.D., *Science* (1999) 283, 1148.

Colville A.A.; Geller S., *Acta Crystallogr.* (1971), *B27*, 2311.

Dance I.G.; Garbutt R.G.; Craig D.C.; Scudder M.L., *Inorg. Chem.* (1987), 26, 4057.

Férey G.; Cheetham A.K., *Science* (1999) 283, 1125.

Férey G., *J. Solid State Chem.* (2000) 152, 37

Férey G., *Angew. Chem. Int. Ed.* (2003) 142, 2576.

Férey G.; Serre C.; Mellot C.; Millange F.; Dutour J.; Surblé S.; Margiolaki I., *Angew. Chem. Intl. Ed.* (2004) 43, 6296.

Férey G.; Serre C.; Mellot C.; Millange F.; Dutour J.; Surblé S.; Margiolaki I., *Science* (2005) 309, 2040.

Férey G.; Millange F.; Morcrette M.; Serre C.; Doublet M.L.; Grenèche J.M.; Tarascon J.-M., *Angew. Chem. Intl. Ed.* (2007) 46, 3259.

François M.; Junod Yvon K.; Hewat A.W.; Capponi J.J.; Strobel P.; Marezio M.; Fischer P., *Solid State Comm.* (1988), 66, 1117.

Genouel R.; Michel C.; Nguyen N.; Hervieu M.; Raveau B., *J. Solid State Chem.* (1995) 115, 469.

Glazer A.M., *Acta Crystallogr.* (1972), *B28*, 3384.

Graab M.; Truckhan M.; Maurer S.; Gummaradju R.; Müller U., *Micro. Mes. Mater.* (2012) 157, 131.

Khilborg L.; Fernandez M.; Laligant Y.; Sundberg M., *Chemica Scripta* (1988) 28, 71.

Lacorre P.; Pannetier J.; Avendunck F.; Hoppe R.; Férey G., *J. Solid State Chem.* (1989) 79, 1.

Li H.; Laine A.; O'Keeffe M.; Yaghi O.M., *Science* (1999) 283, 1145.

Li H.; Eddaoudi M.; Laine A.; O'Keeffe M.; Yaghi O.M.`, *J. Am. Chem. Soc.* (1999) 121, 6096.

Liu Y.-L.; Kravtsov V.C.; Larse R.; Eddaoudi M., *Chem. Comm.* (2006) 1488.

Magnéli A., *Acta Chem. Scand.* (1948) 2, 501.

Magnéli A., *Arkiv für Kemi.* (1949a) 1, 213.

Magnéli A., *Arkiv für Kemi.* (1949b) 1, 269.

Michel C.; Er Rakho L.; Raveau B., *Mater. Res. Bull.* (1985) 20, 667.

Minder W., *Zeitsch. Kristall. Kristall. Phys. Kristall. Chem.* (1937) 96, 15.

Morss L.R., *J. Inorg. Nuclear Chem.* (1974) 36, 3876.

O'Keeffe M., *Acta Crystallogr.* (1992), A48, 879.

Poeppelmeier K.R.; Leonowicz M.Z.; Longo J.M., *J. Solid State Chem.* (1982) 44, 89.

Ross C.R.; Bernstein L.R.; Waychunas G.A., *Amer. Mineral.* (1988) 37, 657.

Tranchemontagne D.J.; Mendoza-Cortes J.L.; O'Keeffe M.; Yaghi O.M., *Chem. Soc. Rev.* (2009) 38, 11506.

Wang C.; Li Y.; Bu X.; Zheng N.; Zivkovic O.; Yang C.S.; Feng P., *J. Am. Chem. Soc.* (2001) 123, 1213.

Yaghi O.M.; Sun Z.; Richardson D.A.; Groy T.L., *J. Am. Chem. Soc.* (1994) 116, 807.

Zhang J.-P.; Chen X.-M., *Chem. Comm.* (2006) 1689.

Zheng N.; Bu X.; Wang B.; Feng P., *Science* (2002) 298, 2366.

Index

www.ingramcontent.com/pod-product-compliance
Lightning Source LLC
Chambersburg PA
CBHW081059220326

41598CB00038B/7153